Atomic Absorption Spectroscopy

Atomic Absorption Spectroscopy

Editor

Federico Noguerra

scitus
academics

Atomic Absorption Spectroscopy
Edited by **Federico Noguerra**

ISBN: 978-1-68117-460-0
Library of Congress Control Number: 2016942819

SCITUS Academics LLC,
www.scitusacademics.com
Box No. 4766, 616 Corporate Way,
Suite 2, Valley Cottage,
NY 10989

Notice

Printed in United States of America on Acid Free Paper

Preface

Atomic absorption spectroscopy (AAS) is a spectroanalytical method for the quantitative determination of chemical elements by means of optical radiation (light) absorption by gaseous free atoms. Atomic absorption is so sensitive that it is possible to measure parts per billion grams of a sample. The technology uses the light wavelengths that an element absorbs in particular. They correspond to the energy needed to promote electrons from one energy level to another at a higher energy level. The atomic absorption spectrometry in different chemistry fields is widely used. Examination of metals in biological fluids such as blood and urine, monitoring of the environment-for example, determining the levels of various elements in rivers, seawater, drinking water, fruit, air, oil, and wine and beverages. Minute quantities of a catalyst (usually a metal) used in the process can sometimes be found in some pharmaceutical production processes in the final product. The quantity of catalyst available can be determined with AAS. Many raw materials are tested and AAS is used extensively to verify, for example, that the main elements are present and that toxic impurities are lower than specified. The level of plumb in concrete, which is a major component of calcium, should be low because it is toxic. The quantity of metals such as gold in rocks can be determined using AAS to determine whether the rocks used to extract gold are worth mining. This book is intended to cover all major topics needed to equip students and practitioners with recent developments in this field.

Table of Contents

CHAPTER 1

Trace Elements Assessment in Human Nails in Eastern Sudan Using Atomic Absorption Spectroscopy

Salwa M. Ibrahim[1], Abdelrazig M. Abdelbagi[2,3], ALi E. Sharf Eldeen[4]

[1]Science College, Sudan University for Science and Technology, Khartoum, Sudan

[2]Physics Department, Science College, Shaqra University, Dawadami, Riyadh, KSA

[3]Faculty of Science, Omdurman Islamic University, Omdurman, Sudan

[4]Faculty of Science, Khartoum University, Khartoum, Sudan

ABSTRACT

In the present work, elements concentration in fingernails samples of volunteers of different ages (males, females) were determined using atomic absorption spectroscopy (AAS) Perkin-Elmer, spectrophotometer. Fingernails samples of different groups were analyzed to determine the trace elements Ca, Mg, Mn, Fe, Ni, Cu, Co, Zn and Pb. Standards materials were prepared for concentration assessment that adjacent to samples from two cities in different location in eastern province of Sudan for elements concentrations finding. In addition, samples of hands fingernails and toenails were analyzed for comparison and method validation. Consequently, the significant levels of elements concentration in nails samples of Jabiat residents compared to Port Sudan resident's area due to soil dispersion are supporting the possibility of external contamination. The data of component matrix and rotated component matrix of varimax normalization using principal component analysis revealed important predictors of nails elements Mg, Mn, Fe, Cu and Zn concentrations in soil of both areas under study. However, the Pb, Ni, Ca, and Co are possible to be indication of different sources associated with environmental contamination. The significant correlation and principal component analysis of the elements of nails concentrations in the two categories supported the probability of different expose environmental contamination.

KEYWORDS

Trace Elements; Nails; Geographical Location; Atomic Absorption Spectrophotometer

1. INTRODUCTION

Since ancient times, the nails growth is used as an indicator of health and physiological imbalances of human. The fingernails and toenails are used as indication of aging, in addition of various properties, such as thickening, thinning, discoloration, splitting, grooves concave and convex shape so that flatness can be used as indicate disease in the body, nutrient deficiencies, drug reaction and poisoning or nail injury. The nails thickness change or loosened and infected with bacteria are illness signed of certain disease [1]. The human nail is permeable than skin and the composition consists of 7% - 12% of water, so that it is a solid part in body. Mechanical pressure on nails can cause harmful pain; also the nails are affected with stretched, tight and cosmetics. Nails after growth is remain isolated from other metabolic activities in the human body, which is considered as a good reflection of long-term exposure (Takagi et al. 1988). The advantages of nails in elements evaluation are preferable biologic medium because of ease of collection, storage convenience, ease of handling and reproducibility of later analysis results. The nails from various fingers in feet and hands are growth in several weeks of time between formation and clipping and that indicates exposure to elevated concentration contamination integrated over a 2 - 12 month period (Hunter 1990).

The trace elements levels in nails are subject of interest in the biomedical and environmental sciences since recent years. Nails measurement remains the subject of interest as indices for assessing nutritional status, diagnosing diseases, identifying systemic intoxication and environmental exposures. The determination of elements contents in the nails can be considered as an indicator of level in other tissues and that reflect mineral metabolism in the body [2] [3] .

The measurement of elements in nails is used as biomarker for exposure, so that the appropriate selection of these elements is of critical importance for health care management purposes, public health decision making, and primary prevention activities [4] . Significant result may lead to develop a better and easy diagnostic measure about metal toxicity particularly as serving program for data recording. Thus, nails can be considered as an excretory product, the trace element contents of which reflect mineral metabolism in the body and leads to contentious data record for many people leading to many valuable properties [5] [6] . The nails human elements concentrations in east district of Sudan relative to the long term source exposure have not been established yet. In these work elements concentration investigation in nails population has possible risk factors exposure to surface soil, drinking water, types of food and other environmental sources [7] . The purpose of this study was to find out a base line information concerning levels of elements exposure in the nails in our population, since previously no reported data are available in this regard in our area [8] .

2. EXPERIMENTAL METHODS

Groups were categorized on geographical location, types of nails. The group's samples were collected from east province of Sudan from various groups of males and females and the climate supposed to be similar in the two cities. In addition, to the underground drinking water is that relative to same source of expose to elements for both sites. Two cities was selected area relative to different criteria such as soil and food, beside the geographical location area with similar demographic characteristics to the exposed area that is indicate elevated concentrations of these elements in the environment [9] .

2.1. The Study Area

Jabiat is part of the red sea hills area and about 380 km far from main national harbor Port Sudan on the red sea coast. The Jabiat district is the east of the red sea coast in the same region of east of Sudan. The secondary chain of hills is separated by sandy valley in the region, which joins Khaour Ariab and Wadi Amur. The climate in the region from June to September is hot season, but in Jabit is moderate. The January and February is coolest period in the region. There are two types of winds in the area and that from the west or northwest during the summer, which is depending on the season and from north or northeast. The Ariab gold mining in the area is closed to Jabiat about 3 km in circle of 25 km diameter and the mineralization of gossans and massive sulphide in main geological locality of the area. The geological history reflect the formation of plutonic rock and oldest contemporaneous with the volcano sedimentary units. Sulphide mineralization is consists of massive pyrite with particular metals and variable Cu and Zn grades and gold a main possible content of arid part of province, which is surrounding Jabiat [10].

Nails samples from various groups were collected from Port Sudan and Jabiat sites.Hands fingernails and toenails were collected from females and males living in different area and of age ranging between 14 - 40 years. The samples were collected from the groups at interval of 2 to 3 weeks, stored in glass containers and then divided relative to localities, gender, ages and an acquired from two subjects as mention. Nails samples collection may seem a relatively simple task and can be easily being obtained from the individuals and there are no significant spatial and temporal variations [11] . The sets of samples are used as a subject of interest and comparable statistical analysis, as well as supporting the validity and reliability of the collected data.

2.2. Samples Preparation

The nails samples were cleaned manually and all visible dirt on the surface removed, and then was washing with water-acetone solution. In the cleaning process of samples the solution of water 50% and acetone 50% were used to remove all types of exogenous contamination, from fats and lipids to various other organic or inorganic substances as recommended by the International Atomic Energy Agency (IAEA, 1985) [12] . Therefore, the heavy metals do not affect with washing procedures to remove the external contaminants due to strong complex with the groups, which is essentially influenced their concentration in the nails sample (Chen et al., 1999), Mikasa et al. (1988) [13] [14] .

Decomposition of heavy metals in nails is very important in determination of these elements by using a mixture of nitric acid (HNO_3) with hydrogen peroxide (H_2O_2) or nitric acid only (Samanta et al., 1999; Chen et al., 1999; Puchyr et al., 1998).

Firstly, standard solutions of each element were prepared by using an amount of 250 μL that was taken from the stock standard solution (1000 mg/L) in a plastic volumetric flask (25 ml) and full up to the mark with 0.5 Normality of hydrochloric acid solution, thus the intermediate amount of standard solution is (10 mg). Secondly, the standard solutions for measurement were prepared to suitable concentration of each element of Ca and Mg in the sample solution. Calcium and Magnesium stock solutions (100 ppm) were taken in a series of 10 ml of plastic volumetric flack that full up to the marks with 0.5 N HCL solution. The solution of Ca and Mg were folded by strontium to prevent interference and the final concentration of series of different amount of Ca and Mg prepared respectively [15] .

The standards of Mn, Fe, Cu, Ni, Co, Zn, and Pb determined by using a series of different concentration were taken from the intermediate standard solution (10 mg/L). The solutions of the elements were filled up to the marks of 0.5 N HCL, which was similar to the solvent of the sample and that to avoid the physical interference, corresponding to concentration of previous elements respectively.

3. RESULTS AND DISCUSSION

Assessments of elements concentration in nails and standards samples are largely based on instrumental Atomic Absorption Spectroscopy elements identification technique in hands and feet nails used [16] . The results were obtained clearly indicate that nails concentration of heavy metals can be used as an indication of many properties such as heavy metal exposure related to a geographical location and soil, drinking water, type of food and may be some certain diseases [17] . The both Figures 1 and 2 are graphically displayed the main elements contents in the nails of Sudanese's males and females of different ages. Relation of elements concentration of nails in Jabiat area is showed higher concentration of Mn, Mg, Fe, Ni, Cu, Co, Zn and Pb with the comparison of Port Sudan area except for Ca in Figures 3 and 4.

The Port Sudan site is revealed the lowest elemental contamination as compared to the Jabait site, which includes the major sources of this site, may be affected with geographical location, soil dispersion and underground drinking water. The elements concentration in nails groups due to residents areas in the east part of Sudan are highly relative to the natural geographical condition and presence of contamination resulting from living activities, environmental sources and underground drinking water [18] .

Figures 5 and 6 shows elements average concentration of feet nails are compared with hands fingernails for all individuals.

The data of fingernails of feet and hands with comparison of available value and the literature reveals the levels of elements Ca, Mg, Mn, Fe,

Cu, Ni, Co, Zn and Pb in the nails samples of feet higher than in hands [19] . The elements level in nails sample was potentially elevated by the absorption of such elements from soil dispersion and underground drinking water in the area and other environmental factors [20] . The toenails are more affected with soil dispersion, bacteria, human activities and other disease when compared to the hands fingernails. Therefore, the content in trace elements of the human nails are vary with locality sites and difference significantly between fingernails of hands and feet [13] .

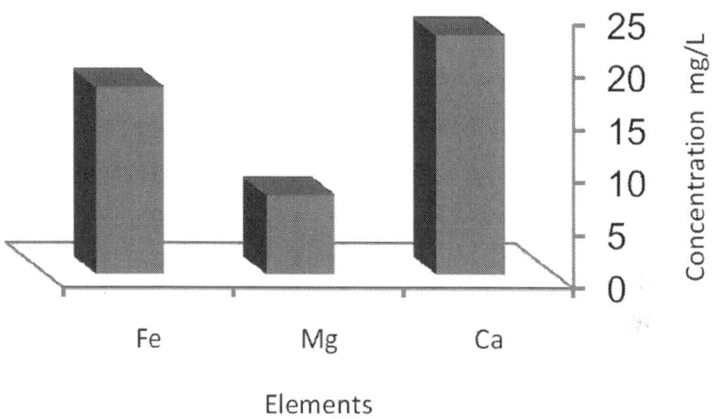

Figure 1. Elements average concentrations of nails.

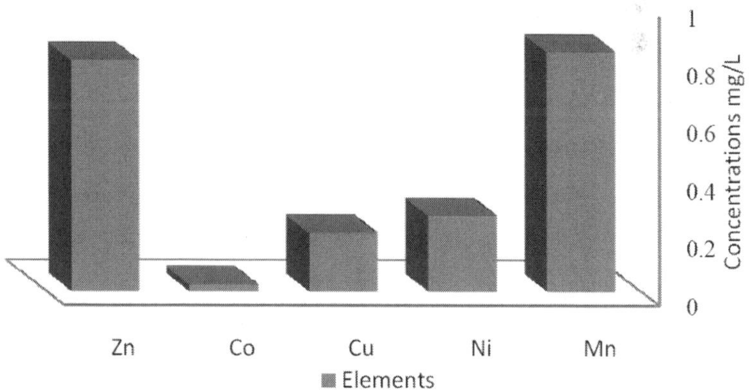

Figure 2. Elements average concentrations of nails.

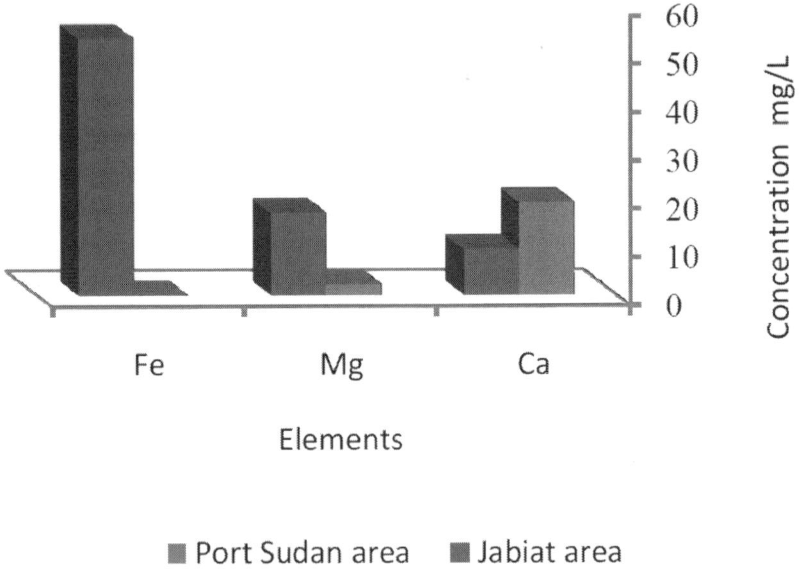

Figure 3. Elements average concentration of nails in different areas.

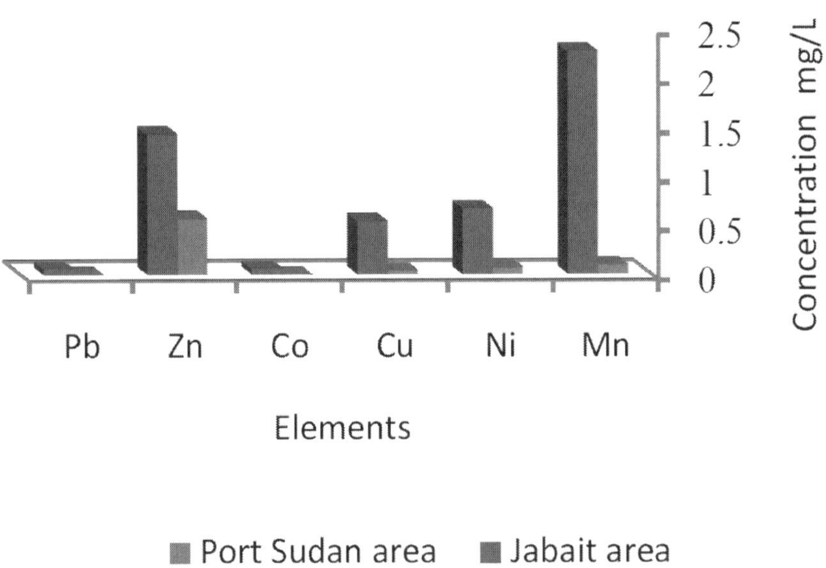

Figure 4. Elements average Cocentration of nails in different areas.

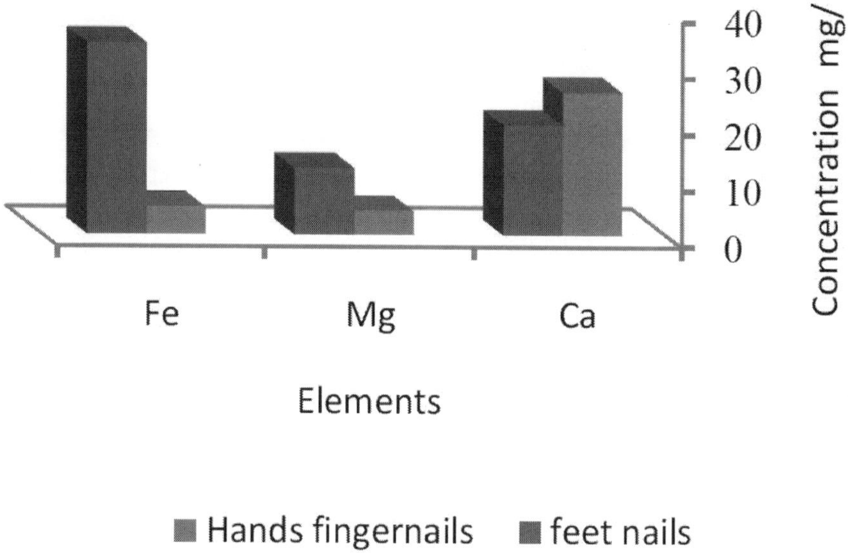

Figure 5. Elements average concentration of nails.

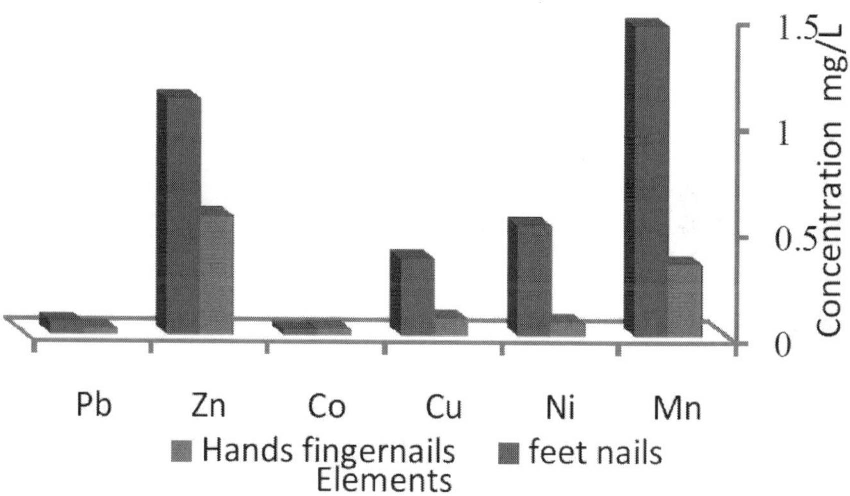

Figure 6. Elements average concentration of nails.

The data in Table 1 shows the correlations of nails elements contents and the marked correlations are significant at $p < 0.05$.

The behaviors of elements concentration in nails were explained by applying the principal component analysis in Table 2.

Therefore, the principal components analysis is categories two groups of elements and the data into three components for component matrix and rotated components matrix with varimax normalization.The relationship in Component (1) of Mn and Mg, Fe, Cu and Zn in nails is possible to be related to geographical exposure and underground drinking water. The data in rotated component matrix with varimax normalization is supporting the high value of Mn and Mg, Fe, Cu and Zn in nails in Component (1).

The high value of Ni and Pb in Component (2) and intermediate value of Ca and Ni in Component (3) are possible to be connected to one source of environmental exposure.

The high value of Pb and negative value of Ca in Component (1) at varimax normalization and the high value of Co as in Component (3) is possible to be indication of different sources associated to environmental exposure for these four elements. The males and females nails elements concentration were illustrated extra of the 82.03% variance in hands fingernails elements concentration and 80.83% variance in feet nails elements concentration as comparative data.

The factors of principal component analysis in Table 3 and statistical data suggest that the higher value of Mn and Mg, Fe, Cu and Zn in fingernails were probably exposed to concentration in soil dispersion related to geographical location and underground drinking water.

Mn is highly correlated with (Mg, Fe, Zn) and moderate correlation with Co. Fe is highly correlated with (Mg, Cu) and moderate correlation with Co. Zn is in highly correlated with (Mg, Fe, Cu) and moderate correlation with Co [20]. The extraction methods of principal components and communalities values indicates high values of Mn, Mg, Fe, Cu and Zn which is excepted one group in factor1, while Ca, Ni and Co of moderate values possible to be connected to one source [21] .

Table 1. Shows the correlation between the elements concentration data of nails samples.

	Ca	Mn	Mg	Fe	Ni	Cu	Co	Zn	Pb
Ca	1.00								
Mn	−0.46	1.00							
Mg	−0.48	1.00	1.00						
Fe	−0.48	1.00	1.00	1.00					
Ni	−0.11	0.25	0.25	0.27	1.00				
Cu	−0.50	0.99	1.00	1.00	0.26	1.00			
Co	−0.57	0.58	0.58	0.56	0.02	0.55	1.00		
Zn	−0.63	0.94	0.94	0.94	0.32	0.95	0.66	1.00	
Pb	−0.35	−0.08	−0.07	−0.07	0.57	−0.04	0.18	0.19	1.00

Table 2. Principal analysis component matrix and rotated component matrix at varimax normalization.

Elements	Component Matrix			Rotated Component Matrix. (Varimax Normalization)		
	Component(1)			Component(2)		
	1	2	3	1	2	3
Ca	−0.626	−0.314	0.553	−0.294	−0.828	−0.155
Mg	0.971	−0.177	0.148	0.974	0.219	0.020
Mn	0.977	−0.166	0.130	0.970	0.240	0.024
Fe	0.974	−0.163	0.153	0.974	0.220	0.035
Ni	0.309	0.707	0.578	0.283	−0.155	0.909
Cu	0.975	−0.140	0.138	0.964	0.240	0.051
Co	0.686	0.078	−0.530	0.419	0.762	−0.044
Zn	0.981	0.072	−0.013	0.861	0.437	0.137
Pb	0.096	0.949	−0.106	−0.205	0.418	0.839

Table 3. Principal componentanalysis factors.

Elements	Factor 1	Factor 2
Ca	0.391435	0.490213
Mn	0.943411	0.974916
Mg	0.953901	0.981555
Fe	0.947868	0.974575
Ni	0.095540	0.596035
Cu	0.950151	0.969728
Co	0.470470	0.476549
Zn	0.962438	0.967646
Pb	0.009136	0.910062

4. CONCLUSION

Certainly, this is the first time to determined trace elements concentration in the nails of males and females in eastern Sudan related to geographical location, types of fingernails and environmental sources in the area. The correlation coefficient and principal component analysis are confirmed that the three groups of elements categorized due to the types of sources .However, these elements can be accumulated by direct external contamination related to environmental sources such as geographical location or indirect exposure due to types of food or drinking water [22] . Nails elements concentration in Port Sudan are more strongly coinciding variation with Jabiat, which is strong likelihood of significant external contamination linked to geographical location and underground drinking water and type of food, and activities in the cities [1] . Accordingly, the hands and feet fingernails elements concentration are differed probably due to external contamination exposure. On this regards, the statistical assessment of nails measurement is to identify potential problems as indicator for possible serious health effects and environmental contamination.

REFERENCES

1. Suhonen, R.E., Dawber, R. and Ellis, D.H. (1999) Fungal Infection of the Skin, Hair and Nails. Martin Dunitz, London.

2. Kanabrocki, E.L., Kanabrocki, J.A., Greco, J., Kaplan, E. and Oester, Y.T. (1979) Instrumental Analysis of Trace Elements in Thumbnails of Human Subjects. Science of the Total Environment, 13, 131-140.

3. Abdulrahman, F.I., Akan, J.C., Chellube, Z.M. and Waziri, M. (2012) Levels of Heavy Metals in Human Hair and Nail Samples from Maiduguri Metropolis, Borno State, Nigeria. World Environment, 2, 81-89.

4. Garland, M., Morris, J.S., Rosner, B.A., Stamfer, M.J., Spate, V.L., Baskett, C.J., Willett, W.C., Hunter, D.J. (1993) Toenail Trace Element Levels as Biomarkers: Reproducibility over a 6-Year Period. Cancer Epidemiology Biomarkers Preview, 2, 493-497.

5. Wilhem, M. and Hafner, D. (1991) Monitoring of Cadmium, Copper, Lead and Zinc Status in Young Children Using Toenails: Comparison with Scalp Hair. Science of Total Environmental, 103, 199-207.

6. Peters, K., Gammelgaard, B. and Menne, T. (1991) Nickel Concentrations in Fingernails as a Measure of Occupational Exposure to Nickel. Contact Dermatitis, 25, 237-241.

7. Oluwole, A.F., Ojo, J.O., Durosinmi, M.A., Asubiojo, O.I., Akanle, O.A., Spyrou, N.M. and Filby, R.H. (1994) Elemental Composition of Head Hair and Fingernails of Some Nigerian Subjects. Biological Trace Elements Research, 43, 443-452.

8. Hinwood, A.L. (2003) Hair and Toenail Arsenic Concentrations of Residents Living in Areas with, High Environmental Arsenic Concentrations. Environmental Health Perspectives, 111, 187-193.

9. Biswas, S.K., Abdullah, M., Akhter, S., Tarafdar, S.A., Khaliquzzaman, M. and Khan, A.H. (1984) Trace Elements in Human Fingernails: Measurement by Proton-Induced X-Ray Emission. Journal of Radioanalytical Nuclear Chemistry, 82, 111-124.

10. Sam, A.K. and Al-Geed, A.M.M. (2000) Radiological Evaluation of Gold Mining Activities in Ariab (Eastern Sudan). Radiation Protection Dosimetery (Technical Note), 88, 335-340.

11. Rakib, M.A., et al., (2013) Arsenic Content in Inactive Tissue: Human Hair and Nail. Journal of Scientific Research & Reports, 2, 522-535.

12. Vance, D.E., et al., (1988) Trace Element Content in Fingernails and Hair of a Non-Industrialized US Control Population. Biological Trace Element Research, 17, 109-121.

13. Mehra, R. and Juneja, M. (2005) Elements in Scalp Hair and Nails Indicating Metal Body Buren in Polluted Environment. Journal of Scientific and Industrial Research, 64, 119-124.

14. Rao, K.S., Balaji, T. and PrasadaRao, T. (2002) Determination of Iron, Cobalt, Nickel, Manganese, Zinc, Copper, Cadmium and Lead in Human Hair by Inductively Coupled Plasma Atomic Emission Spectrometry. Spectrochimica Acta Part B, 57, 1333-1338.

15. Biswas, S.K., Abdullah, M., Akhter, S., Tarafdor, S.A., Khaliquzzaman, M. and Khan, A.H. (1984) Trace Element Contents in Fingernails and Hair a Human. Journal of Radioanalytical Nuclear Chemistry, 82, 111-124.

16. Rodushkin, I. and Axelsson, M.D. (2000) Application of Double Focusing Sector Field ICP-MS for Multi-Elemental Characterization of Human Hair and Nails. Part II, A Study of the Inhabitants of Northern Sweden. Science Total Environmental, 262, 21-36.

17. Samanta, G., et al. (2004) Arsenic and Other Elements in Hair, Nails, and Skin-Scales of Arsenic Victims in West Bengal, India. Science of the Total Environment Elsevier, 326, 33-47.

18. Muralidharan, L. (2013) Nail as a Diagnostic Tool to Detect Heavy Metal Accumulation in Man Residing in Mumbai City and Their Impact on General Health. International Journal of Advanced Research, 1, 39-44.

19. Othman, I. and Spyrou, N.M. (1980) The Abundance of Some Elements in Hair and Nail from the Machakos District of Kenya. Science of total Environmental, 16, 267-278.

20. Chowdary, K., Ehmann, W.D., Rengan, K. and Markesbery, W.R. (1995) Trace Elements Correlation with Age and Sex in Human Nails. Journal of Radioanalytical Nuclear Chemistry, 195, 51-66.

21. Gammelgaard, B., Peters, K. and Menne, T. (1991) Reference Values for the Nickel Concentration in Human Finger nails. Journal of Trace Elements and Electrolytes in Health and Disease, 5, 121-123.

22. Karagas, R.M., Tosteson, T.D. and Blum, J. (2000) Measurement of Low Levels of Arsenic Exposure: A Comparison of Water and Toenail Concentrations. American Journal of Epidemiology, 152, 84-90.

CHAPTER 2

Branched Polyamines Functionalized with Proposed Reaction Pathwaysbased on ^{1}H-NMR, Atomic Absorption and IR Spectroscopies

Vicente Cervantes-Mejía, Elizabeth Baca-Solis, Judith Caballero-Jiménez, Rosario Merino-García, Jesús Cruz-Gatica, Gabriela Moreno-Martínez, Yasmi Reyes-Ortega

Chemical Center, Sciences Institute, Autonomous University of Puebla, Puebla, México

ABSTRACT

Three novel branched polyamines N,N,N',N'-tetrakis-[3((pyridine-2-methyl)-amine) propyl]-1,4- butanediamine (1), N,N,N',N'-tetrakis-[N-((2-methylpyridine)ethyl)propanamide]ethylenediamine (2) and N,N,N' ,N'-tetrakis-[3((2-hidroxibenziliden)-amine)propyl]-1,4-butanediamine (3), were syn- thesized starting from 2-pyridinecarboxaldeyde with DAB-Am-4 for 1, PAMAM G0 for 2 and from salicylaldehyde with DAB-Am-4 for 3. The pathway reactions have been proposed by ^{1}H-NMR, IR and Atomic Absorption Spectroscopy. The optimal reaction time was set by IR spectroscopy following aldehyde $\bar{\nu}_{(C=O)st}$ peak modification. 1 and 2 were obtained as both hydrochlorides and as free amines and 3 only as free imine. These polyamines were characterized by UV-Vis, IR, ^{1}H-NMR and ^{13}C-NMR and Mass Spectrometry.

KEYWORDS

Branched Polyamines, Functionalization Reactions, IR, NMR, Atomic Absorption Spectroscopy

1. INTRODUCTION

Since the synthesis of the first branched polyamines in the late 1970s, these repetitively three-dimensional polymers have provided a rich seam of research due to their wide range of applications [1] [2] . The branched polymers can be used as low-dielectric materials [3] , as templates for the growth of single-wall carbon nanotubes [4] , as catalysts [5] [6] , as biosensors [7] , as optoelectronic devices [8] [9] as well as in biological applications [10] , in magnetic resonance imaging [11] [12] , in drug delivery [13] , and in coordination chemistry [14] , among others.

Several methods for the synthesis and modification of linear polyamines are available and depend on the number and type of N atoms, the hydrocarbonated chain length between the amine groups, and the conforma- tional changes when cyclic units are introduced [15] [16] . For tetra branched molecules, it is possible to tune their structure, size, shape and solubility; however, their modification through standard condensation reactions implies a complicated mixture of mono-, bi- and tri-substituted side products. Then, the functionalization has been proven to be successful method to modify the outer sphere of preformed branched structures. In this research we are reporting three new functionalized branched molecules, N,N,N',N'-tetrakis-[3((pyridine-2-methyl)-amine)propyl]-1,4-butanediamine 1, N,N,N',N'-tetrakis-[N-((2-methylpyridine)ethyl)propanamide]ethylenediamine 2 and N,N,N',N'-

tetrakis-[3((2-hidroxibenziliden)-amine)propyl]-1,4-butanediamine 3. We introduced a new strategy for the synthesis of 1 and 2 by modifying the DAB-Am-4 and PAMAM G0 cores by using an inexpensive catalyst and mild reaction conditions. Compound 3 was synthesized using the reported reaction of obtaining of imines. IR spectroscopy and thin layer chromatography (TLC) were used to follow the reaction and to obtain fully functionalized branched molecules with the best yields in the optimal reaction times. Combination of NMR and Atomic Absorption spectroscopies were used to propose the pathway reactions of 1 and 2. Our interest lies in the introduction of the aromatic rings with electron-density-donor atoms to make these molecules ideal candidates to obtain coordination compounds with potential applications in the future.

2. MATERIALS AND METHODS

2.1. Reagents and Instrumentation

First, all the reagents and solvents were purchased from Aldrich and used without further purification. Melting points were obtained with PF-300 SEV equipment. UV/Vis spectra were obtained with a Shimadzu UV-3100S spectrophotometer on MeOH, $CHCl_3$ solutions of ca 8.8 ´ 10^{-5} M for 1 and 1.84 ´ 10^{-4} M for 2 as hydrochlorides. As free amines, in MeOH, $CHCl_3$ solutions using ca of 1.16 ´ 10^{-3} M/7.57 ´ 10^{-5} M for 1 and 6.85 ´ 10^{-5} M/1.60 ´ 10^{-5} M for 2, using CH_2Cl_2 and CH_3OH for 3 at ca 10^{-5} M. UV-Vis 1-3 data are reported as: wavelength, l_{max} (nm), molar absorptivity, ε (cm^{-1}·mol), transition energy gap, ΔE (cm^{-1}). IR spectra were recorded with a Nicolet Magna-IR 750 spectrophotometer between 4000 cm^{-1} and 400 cm^{-1} using KBr pellets. The ^{1}H-NMR and ^{13}C-NMR were recorded with a Bruker Avance III 500 MHz spectrometer for 1 and 2, and with a JEOL Eclipse-400 spectrometer for 3, in $CDCl_3$ solutions and using TMS as reference for all samples. ^{1}H-NMR spectra were recorded between 0 and 10 ppm with acquisition time of 2 sec, 16 repetitions. Chemical shifts are reported as δ part per million (ppm) values relative to TMS; s = singlet, d = doublet, t = triplet, m = multiplet. The mass spectra were recorded in a JEOL MStation JMS-700 spectrometer using a FAB^{+} technique. A GBC 932 atomic absorption spectrometer was used to quantify the Zn^{2+} concentrations by the flame method. The hollow cathode lamp was operated at 5.0 mA, with λ = 213.9 nm, a slit of 0.5 nm and burner height of 13.5 mm, the airflow and acetylene rates were used as recommended by the manufacturer and a background correction was carried out with a deuterium lamp. Aqueous solutions of $Zn(C_2H_3O_2)_2 \cdot H_2O$ at 0.5, 1.0, 1.5 mg/L concentrations were used to obtain a calibration curve.

2.2. General Synthetic Procedure

General synthetic procedures to obtain 1 and 2: A mixture of acetic acid (0.658 mL, 11.49 mmol), Zn^0 (657.8 mg, 10.06 mmol) and DAB-Am-4 (320.0 mg, 1 mmol) for 1 or PAMAM G0 (516.5 mg, 1 mmol) for 2 in 8 mL of MeOH were stirred under reflux. Then, 2-pyridinecarboxaldehyde (0.38 mL, 6 mmol), in 3 mL of MeOH was added dropwise over 1.5 hours. The mixture was refluxed for 6 hours for 1 and 72 hours for 2, cooled to room temperature and filtered. Then large amounts of gaseous HCl were bubbled until a white precipitate appeared. The solid was decanted and washed three times with MeOH and Et_2O. The free products are obtained by treating the hydrochlorides with 10 mL of a 1 M NaOH solution, extracted with chloroform and dried.

General synthetic procedure to obtain 3: A mixture of DAB-Am4 (1.49 mmol) with 1 mL of a 5.98 mM solution of salicylaldehyde in MeOH was stirred in a flask. The reaction mixture changed immediately from yellow to orange when salicylaldehyde was added. Then, 1 mL of salicylaldehyde aliquots were added at 0.5 hrs, 1.5 hrs and 4 hrs from the starting reaction time. The reaction was followed by silica TLC, IR and ^{1}H-NMR spectroscopy. After 24 hrs 3 was extracted with $CHCl_3$ from the reaction mixture, dried and filtrated.

N,N,N',N'-tetrakis-[3((pyridine-2-methyl)-amine) propyl]-1,4-butanediamine (1). 1 was obtained as yellow oil (yield: 84.1%). UV-Vis/CH_3OH: $l_{max}/\varepsilon/\Delta E$ ($nm/cm^{-1}mol/cm^{-1}$): 261/9815/38314.2; UV-Vis/$CHCl_3$ $l_{max}/\varepsilon/\Delta E$ ($nm/cm^{-1}mol/cm^{-1}$): 262/573/38167.9; Selected IR data/KBr $\bar{\nu}$ (cm^{-1}): $\bar{\nu}_{O\text{-}H}$ 3416 (br), $\bar{\nu}_{N\text{-}H}$ 3257 (vs), $\bar{\nu}_{C\text{-}H}$ 2934 (s), $\bar{\nu}_{C\text{-}H}$

2817 (m), $\bar{\nu}_{C\text{-}N}$ 1649 (w), $\bar{\nu}_{C=C}$ 1593 (m), $\bar{\nu}_{N\text{-}H}$ 1567 (m), $\bar{\nu}_{C\text{-}Har}$ 1470 (m), $\bar{\nu}_{C\text{-}N}$ 1116 (w), $\bar{\nu}_{C\text{-}H}$ 758 (m); ^{1}H-NMR (500 MHz, $CDCl_3$) δ (ppm): 8.54 (d, 4H), 7.62 (t, 4H), 7.299 (s, 4H), 7.14 (t, 4H), 3.88 (s, 8H), 2.66 (t, 8H), 2.46 (t, 8H), 2.38 (s, 4H), 1.66 (q, 8H), 1.37 (s, 4H), ^{13}C-NMR (500 MHz, $CDCl_3$) δ (ppm): 159.90, 149.23, 136.39, 122.22, 121.84, 55.29, 54.01, 52.22, 48.17, 27.47, 24.92; MS (FAB^+) for $C_{40}H_{60}N_{10}$ $[M + H]^+$ calculated: 680.99, found: 681.00.

N,N,N',N'-tetrakis-[N-((2-methylpyridine)ethyl)propanamide] ethylendiamine (2). 2 was obtained as a yellow oil (yield: 60%). UV-Vis/CH_3OH: l_{max}/ε/ΔE (nm/cm^{-1}mol/cm^{-1}): 205.1/48757/48756.7, 207.9/48100/ 48100, 211.4/47304/47303.7, 218.3/47304/45808.5, 257.3/38865/38865.1, 261.5/38241/38240.9, 268.1/37300/ 37299.5; UV-Vis/$CHCl_3$ l_{max}/ε/ΔE (nm/cm^{-1}mol/cm^{-1}): 252.5/39604/39604, 263.4/37965/37965.1, 270.6/36955/ 36954.9, 309.8/32279/32278.9; IR/KBr $\bar{\nu}_{max}$ (cm^{-1}): $\bar{\nu}_{N\text{-}H}$ 3278 (vs), $\bar{\nu}_{C\text{-}H}$ 2927 (m), $\bar{\nu}_{C=O}$ 1647 (vs), $\bar{\nu}_{C=C}$ 1593 (s), $\bar{\nu}_{N\text{-}H}$ 1568 (m), $\bar{\nu}_{C\text{-}H}$ 761 (m); ^{1}H-NMR (400 MHz, $CDCl_3$): δ ppm: 8.52 (s, 4H), 8.65 - 7.16 (m, 12H), 3.87 (d, 8H), 3.34 (m, 8H), 2.77 (m, 8H), 2.71 (m, 8H), 2.62 (s, 4H), 2.29 (m, 8H); ^{13}C-NMR (400 MHz, $CDCl_3$) δ (ppm): 173.26, 159.55, 149.43, 136.87, 122.61, 122.30, 54.8, 52.12, 50.77, 48.68, 39.46, 34.37; MS (FAB^+) for $C_{46}H_{68}N_{14}O_4$: $[M]^+$ calculated: 881.10, found: 881.00.

N,N,N',N'-tetrakis-(3((2-hidroxibenziliden)-amine)propyl)-1,4-butanediamine (3). 3 was obtained as a highly viscous yellow liquid (yield: 83.0%). UV-Vis/CH_3OH: l_{max}/ε/ΔE (nm/cm^{-1}mol/cm^{-1}): 216.9/45426.4/ 46053.6, 252.61/20013.5/39544.8, 269.5/8926.9/37065.1, 310.8/7415.1/32150.1, 393.5/2879.6/25391.5, 515.1/ 503.9/19396.2; UV-Vis/$CHCl_3$: l_{max}/ε/ΔE (nm/cm^{-1}mol/cm^{-1}): 248.9/30429.4/40126.3, 260.2/33836.9/38394.2, 271.3/10380.9/36824.6, 317.7/13550.6/31439.7, 406.0/554.7/24605.9; IR/KBr $\bar{\nu}_{max}$ (cm^{-1}): $\bar{\nu}_{O\text{-}H}$ 3060 (vb), $\bar{\nu}_{C=N}$ 1633 (s); ^{1}H-NMR (400 MHz, $CDCl_3$) δ (ppm): 1.39 (s, 4H), 1.79 (m, 8H), 2.38 (s, 4H), 2.49 (t, 8H), 3.59 (t, 8H), 7.30 - 6.81 (m, 16H), 8.31 (s, 4H), ^{13}C-NMR (400 MHz, $CDCl_3$) δ (ppm): 25.08, 28.40, 51.33, 53.90, 57.35, 116.94, 118.38, 118.72, 131.07, 132.02, 161.25, 164.82; MS (FAB^+) for $C_{44}H_{56}N_6O_4$: $[M]^+$ calculated: 732.88 found: 733.00.

3. RESULTS AND DISCUSSION

3.1. Synthesis

A previously reported methodology from our research group was used to synthesize compounds 1 and 2 [17] (Figure 1). This methodology was originally proposed for reactions between linear chain polyamines and 2-pyridinecarboxaldehyde and adapted for the functionalization of the DAB-Am4 and PAMAM cores leading to 1 and 2 in good yields. On the other hand, compound 3 showed in Figure 1 was obtained using a reported methodology from the literature to synthesize branched imines [18] .

The pathway reaction to obtain 1 and 2 was proposed based on the Zn redox potential quantification, IR and NMR spectroscopic studies (Scheme 1), whose discusses are given in the section corresponding.

We corroborated that Zn oxidation potential to produce H^- is feasible at the synthesis reaction conditions. For that purpose a mixture of Zn^0, acetic acid and methanol were stirred during 7.5 hours. Subsequently, the reaction mixture was filtered, and subsequently diluted with deionized water. The initial and final concentration of Zn^{2+} were quantified by atomic absorption spectrometry. The obtained values were $[Zn^{2+}]_i = 0.0$ mg/L and the $[Zn^{2+}]_f = 38{,}125$ mg/L, for the initial and final concentrations, respectively. At 343 K (70°C) and at pH = 2.78 Zn^{2+}/Zn the oxidation-reduction potential was quantified using equation (1) [19] .

$$E = E_0 - \frac{RT}{\nu F}[\ln Q] \quad (1)$$

where R = ideal gas constant, T = 343 K, ν = 2 transferred electrons, F = Faraday constant and $Q = [Red_A]^{a'}[Ox_B]^{b'}/[Ox_A]^{a}[Red_B]^{b}$. The redox potential value was of E = −0.85 V and $\Delta G = -39.10$ kcal·mol^{-1}, it implies that the Zn oxidation reaction is favored. These results showed that with these reaction conditions, the Zn^0 oxidation in acid media allowed for hydride formation, H^-. The H^- ions are able to reduce fully the pyridinecarboxaldehyde carbonyl. With this in mind, the proposed reaction mechanism is shown in the Scheme 1(a).

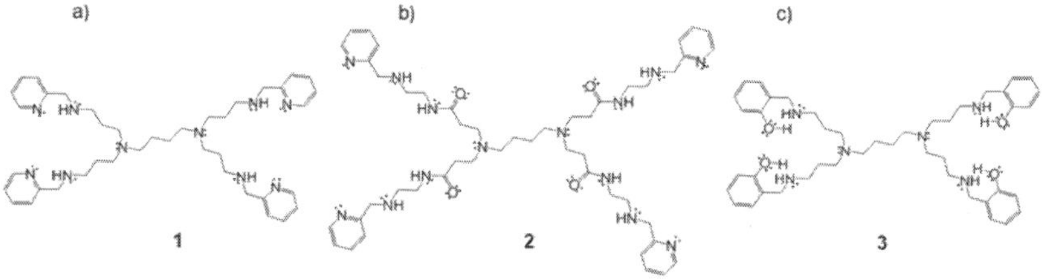

Figure 1. Chemical structures of compounds (a) 1; (b) 2; and (c) 3. The H atoms were omitted for clarity.

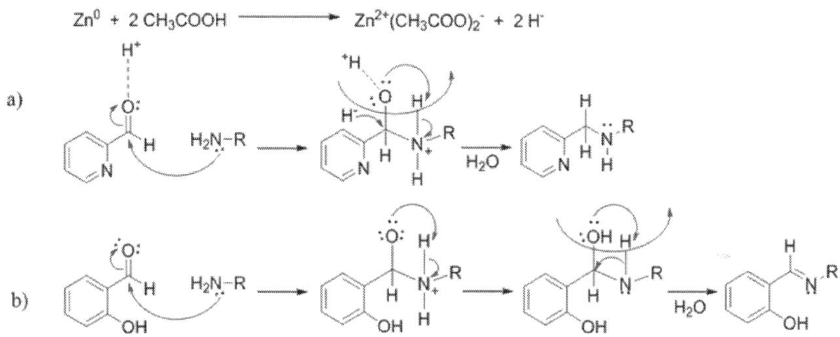

Scheme 1. (a) Proposed reaction pathway for the synthesis of 1 and 2; and (b) Reaction pathway for 3 [17] [18].

In the first step there is a nucleophilic attack of the amino group contained in the DAB-Am4 or PAMAM-G0 molecules due the carbonyl π-electrons polarization. In the second step a concerted mechanism involving the H^- ions is carried out, which avoids the imine formation. The H^- ion is highly reactive and does not allow the formation of double bonds and therefore the formation of imines.

On the other hand, the current tendency in the conventional condensation reactions to obtain imine [18] is the optimization and reduction of the solvents, energy and secondary reagents used for the synthesis. This is known as "green chemistry" [20] . The obtaining of 3 without using solvents was not possible since the mixture reaction showed a high viscosity and a null reactants mixture results; however the use of solvents in small amounts lead to obtain 3 in good yields. The classic reaction mechanism for 3 is given in the Scheme 1(b).

It is noteworthy that although the salicylaldehyde has an electron-donor hydroxo group in an ortho position with respect to the aldehyde group fast protonation of the oxygen atom was carried out. This shows that when H^- ions are not in the reaction medium, imines compounds will always be obtained [18] . Therefore we conclude that when a strong nucleophile (as the hydride anion) is found in the media reaction the amine formation must be favored.

3.2. IR Spectroscopy

IR spectra of 1-3 showed the vibrations typical of the materials containing DAB-Am4, PAMAM-G0, 2-pyridinecarboxaldehyde and salicylaldehyde (Supporting Information Figure S1) [21] - [26] . The optimal reaction times for obtaining 1-3 were set by taking samples at different reaction times, observing the change in the spectra. Figure 2 shows the spectra of (a) the starting materials and (b) to (e) the reactions evolution for 1-3 at different reaction times in the zone between 2200 cm^{-1} and 1100 cm^{-1}.

The band around 1715 cm^{-1} is assigned to the characteristic vibration $\bar{\nu}_{C=O}$ st of the 2-pyridinecarboxalde- hyde, and at 1664 cm^{-1} is the characteristic vibration $\bar{\nu}_{C=O}$ from the salicylaldehyde carbonyl group [22] [25] - [27] . Figure 2(i) and Figure 2(iii) show the IR spectra evolution, at different times of reaction to obtain 1 and 3. For compound 2 the monitoring through the infrared spectra (Figure 2(ii)) was complicated due to the presence of the carbonyl in the amide group, which is very close to the carbonyl signal from the starting material. However, the carbonyl bands in the final spectrum were fitted with Gaussians showing that the carbonyl signal corresponding to 2-pyridinecarboxaldehyde disappeared completely.

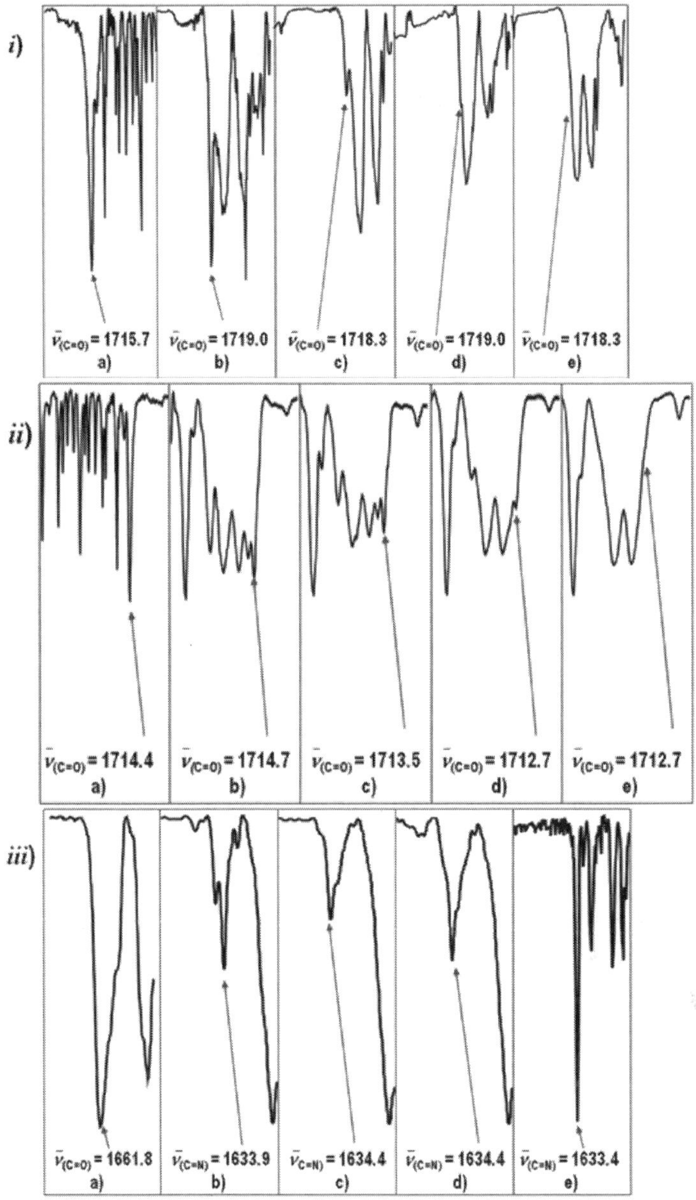

Figure 2. IR spectra (cm^{-1}) showing the region for (i) Com- pound 1, (b) synthesis reaction at 1 hour, (c) 3 hours, (d) 5 hours and (e) 6 hours; (ii) Compound 2: (a) 2-pyridinecarboxaldehyde, (b) synthesis reaction at 3 hours, (c) 7 hours, (d) 24 hours and (e) 72 hours; and (iii) Compound 3: (a) salicylaldehyde, (b) shyntesis re- action at 1 hour, (c) 2 hours, (d) 4 hours and (e) 24 hours.

The optimal time for the reaction for three compounds was also corroborated by TLC. Finally, it is noteworthy that the energy position of the OH band of 3 implies two characteristics: 1) the hydroxyl group is conjugated with amino group through the aromatic ring; 2) the equilibrium phenol-imine and keto-amine tautomeric forms are shifted to phenol-imine form (Supporting Information in Figure S2) [27] - [29] . Table 1 summarizes the most important vibration frequencies found in the infrared spectra assigned according the literature [27] - [29] .

3.3. UV-Vis Spectroscopy

The UV-Vis spectra in MeOH show transitions with λ_{max} = 258.5 nm for 1, λ_{max} = 205.1 nm, 207.9 nm and 257.3 nm for 2 and λ_{max} = 216 nm, 252.6 nm, 269.5 nm, 310.7 nm, 393.4 nm and 515 nm for 3 [22] [26] [27] [29] [30] .

Through a fitting of the spectra with Gaussians it is possible to know the exact number of electronic transitions contained on each band. For 1, the fitting shows four Gaussians with λ_{max} = 241.6 nm, 258.8 nm, 261.7 nm and 267.9 nm (Supporting Information in Figure S3), which is in agreement with the number and energy of the π-π^* transitions reported for the pyridine chromophore [29] . For 2 (Figure 3(a)), the diverse functional groups lead to a more complex spectrum presenting two broad bands containing seven Gaussians assigned to four n-π^* electronic transitions with λ_{max} = 205.1 nm, 207.9 nm, 211.4 nm and 218.3 nm and three n-π^* transitions with λ_{max} = 257.3 nm, 261.5 nm and 268.1 nm. The spectrum for 3 shows four broad bands containing six Gaussians assigned to two n-π^* electronic transitions with λ_{max} = 216.9 nm and 252.6 nm and four n-π^* transitions with λ_{max} = 269.5 nm, 310.7 nm, 393.4 nm and 515 nm in agreement with the structure of 3 (Supporting Information in Figure S4).

A change of the solvent used in the UV-Vis experiments, from MeOH to $CHCl_3$, had practically no effect on the number and wavelength of the electronic transitions for compound 1. This spectrum showed a band with λ_{max} = 262 nm, which fitted with four Gaussians with λ_{max} = 252.5 nm, 263.4 nm, 270.6 nm and 309.8 nm (Supporting Information in Figure S3). However, compounds 2 and 3 UV-Vis spectra showed a significant broadening and shifting of the bands. For 2 the bands with λ_{max} = 257.3

nm, λ_{max} = 261.5 nm and λ_{max} = 268.1 nm in the spectrum of MeOH are shifted to λ_{max} = 263.4 nm, λ_{max} = 270.6 nm and λ_{max} = 309.7 nm in the spectrum of $CHCl_3$. Figure 3(a) and Figure 3(b) show the UV-Vis spectra of 2 in MeOH and $CHCl_3$ highligting the bathochromic shifting. We attribute this change in the spectra to the stabilization of the n orbitals due to H-bonding formation between the solvent and the amide groups in compound 2 when MeOH is used as the solvent [30] . For 3, the transitions were recorded in CH_3OH and CH_2Cl_2 (Supporting Information in Figure S4). The OM diagram clearly shows (Supporting Information in Figure S4), the stabilization of 3 in CH_2Cl_2 due the intra-hydrogen- bond between the N of the imine group and the hydroxyl proton group adopting the phenol-imine form (Supporting Information in Figure S2); while the spectrum in MeOH solution shows that the methanol-imine intermolecular hydrogen bridges are present and therefore 3 is destabilized. This corroborates the information obtained through the IR spectra for 3 and must be valid for 2 [29] [31] .

3.4. NMR Spectroscopy

All three compounds are obtained as yellow oils very soluble in MeOH, H_2O and $CHCl_3$. The obtaining of the desired compounds is corroborated by the number, chemical shifts and integral peak intensities in the ^{1}H-, ^{13}C-NMR spectra, which are in good agreement with the chemical formulas. The peaks are correctly assigned carrying out HSQC experiments, which correlate with the heteronuclear C and H atoms directly bonded.

The pyridine ring signals in the ^{1}H-NMR spectra are found between 7.14 ppm - 8.54 ppm for 1 (Supporting Information in Figure S5) and between 7.16 ppm - 8.53 ppm for 2, while for 3 the aromatic protons are found between 6.81 ppm - 7.30 ppm (Figure 4(a), Figure 4(b)). The peaks between 3.87 ppm - 1.26 ppm for 1, 3.87 ppm - 2.29 ppm for 2 and 3.59 ppm - 1.39 ppm for 3 are assigned to the aliphatic zone. The aromatic signals are correlated in the ^{13}C-NMR spectra with the peaks between 159.90 ppm - 121.84 ppm for 1, between 154.55 ppm - 122.30 ppm for 2 and between 164.82 ppm - 116.94 ppm for 3 and in the aliphatic zone with the peaks between 55.29 ppm - 24.92 ppm for 1, between 54.74 ppm - 29.71 ppm for 2 and between 57.35 ppm - 25.08 ppm for 3 [32] .

Table 1. Main vibrational bands found for 1-3.

	Frequencies $\bar{\nu}$ (cm^{-1})									
	$\bar{\nu}_{(N-H)st}$	$\bar{\nu}_{(C-H)st\ ar}$	$\bar{\nu}_{(O-H)st}$	$\bar{\nu}_{(C=O)st}$	$\bar{\nu}_{(C=N)st}$	$\bar{\nu}_{(C=C)st}$	$\bar{\nu}_{(N-H)\delta}$	$\bar{\nu}_{(O-H)\delta}$	$\bar{\nu}_{(C-O)st}$	$\bar{\nu}_{(C-N)st}$
Compound **1**	3257	2934 2817	-	-	-	1593	1567	-	-	1116
Compound **2**	3278	2927	-	1647	-	1593	1568	-	-	-
Compound **3**	-	3060	2788	-	1633	1577	-	1282	1207	-

ar = aromatic, al = aliphatic.

To corroborate the reaction mechanism proposed for the synthesis of 1 and 2, the ^{1}H-NMR spectra of 1 were obtained at 1, 3 and 5 min from the starting reaction time showing that there was no double bond formation (Supporting Information in Figure S6); while the characteristic proton and carbon signals in the HSQC spectrum of 2, 3 were very clear (Figure 4(a), Figure 4(b)).

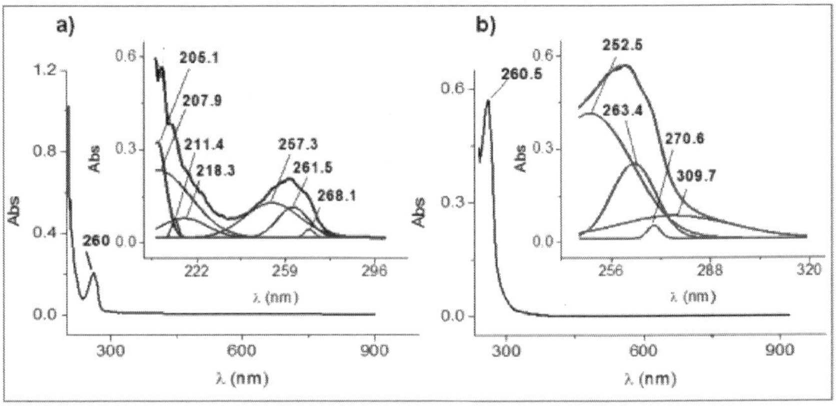

Figure 3. UV-Vis spectra of 2 showing the shifting of the electronic transi- tions wavelengths by varying the solvent from (a) MeOH to (b) $CHCl_3$. The upper right zooms in the 200 - 300 cm^{-1} zone.

Thus, NMR studies corroborated the information obtained by IR about the reaction mechanism for obtaining 1 and 2, which does not imply the imine formation due the presence of a strong nucleophile agent (H^-) which acts fast on the carbonyl carbocation.

Because the molecules contain two perpendicular symmetry axes, only the H^+ and C atoms of one of the branches in the molecules are shown. The spectra exhibit broad signals because of the flexible nature of these molecules, which is more evident in the spectrum of 2, which contains a PAMAM core and therefore more functional groups.

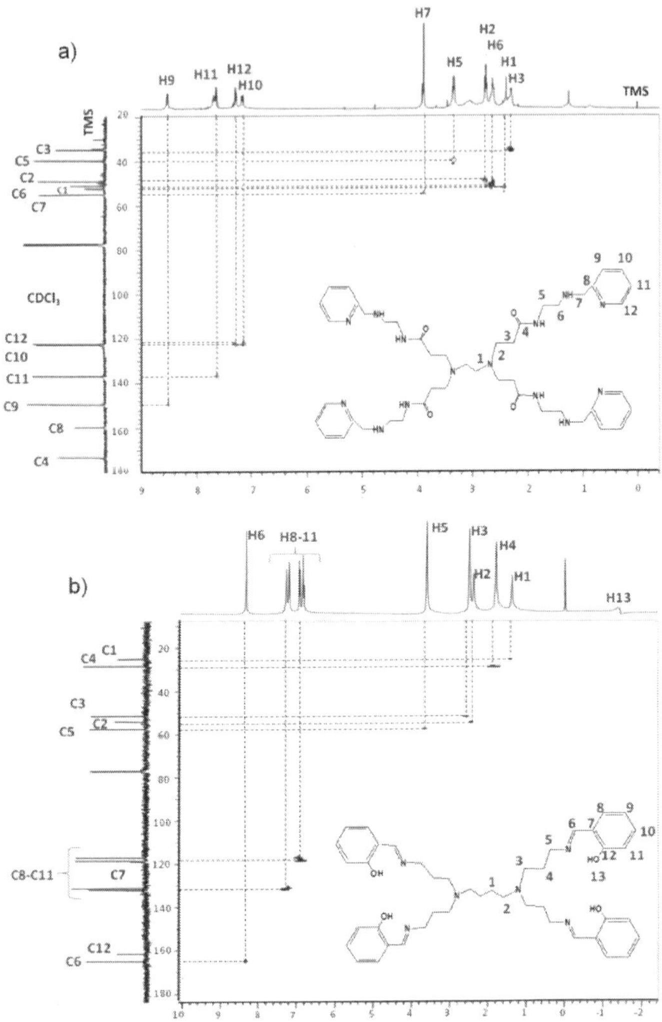

Figure 4. HSQC spectra of (a) 2 and (b) 3, showing the correlation among the protons and carbons present in one branch of the molecules.

4. CONCLUSION

Two different methodologies were used to obtain three fully functionalized branched molecules. The functionalizations were very selective and led to good yields. The classic condensation reactions to obtain only imine- products, were carried out in soft conditions without strong nucleophiles in the reaction medium; while the reductive condensation produces compounds totally functionalized and an unique amine-product. The conditions of the reductive condensation reaction lead to a concerted mechanism which does not allow an imine intermediary and it directs the reaction towards the obtainment of amines. On the other hand, typical reaction conditions for obtaining imine bonds do not produce a nucleophile as the hydride anion, which is unstable respect to the intermediary ammonium ion and reacts forming an amine. The most remarkable conclusion of this work is that the proposed reductive condensation synthesis for amines-carbonyls is a route to obtain amines with high purity and with the best yields, which can be followed by common spectroscopic methods. Finally, the IR, ^{1}H-NMR, TLC, Atomic Absorption, and MS were powerful analytical methods that supported these proposals.

ACKNOWLEDGEMENTS

We thank CONACyT Mexico and the Autonomous University of Puebla for financial support with projects Y-NAT11-I, Y-NAT12-I and Y-NAT13-I, Strengthening of the Doctorate Programs Recognized in PNPC- CONACyT, 2013-2014; we also thank Dra. Margarita Teutli for carrying out the Atomic Absorption quantifications.

REFERENCES

1. Vögtle, F., Buhleier, E.W. and Wehner, W. (1978) Cascade and Nonskid-Chain-Like Syntheses of Molecular Cavity Topologies. Synthesis, 2, 155-158.

2. Bosman, A.W., Janssen, H.M. and Meijer, E.W. (1999) About Dendrimers: Structure, Physical Properties, and Applications. Chemical Reviews, 99, 1665-1688.

3. Lee, B., Park, Y.H., Hwang, Y.T., Oh, W., Yoon, J. and Ree, M. (2005) Ultralow-k Nanoporous Organosilicate Dielectric Films Imprinted with Dendritic Spheres. Nature Materials, 4, 147-150.

4. Amama, P.B., Maschmann, M.R., Fisher, T.S. and Sands, T.D. (2006) Dendrimer-Templated Fe Nanoparcles for the Growth of

Single-Wall Carbon Nanotubes by Plasma-Enhanced CVD. Journal of Physical Chemistry B, 110, 10636- 10644.

5. Twyman, L.J., King, A.S.H. and Martin, I.K. (2002) Catalysis inside Dendrimers. Chemical Society Reviews, 31, 69-82.

6. Kainz, Q.M. and Reiser, O. (2014) Polymer and Dendrimer-Coated Magnetic Nanoparticles as Versatile Supports for Catalyst, Scavengers, and Reagents. Accounts of Chemical Research, 47, 667-667.

7. Mynar, J.L., Lowery, T.J., Wemmer, D.E., Pines, A. and Fréchet, J.M. (2006) Single Quantum Dot-Micelles Coated with Silica Shell as Potentially Non-Cytotoxic Fluorescent Cell Tracers. Journal of the American Chemical Society, 128, 6334-6335.

8. Lo, S.-Ch. and Burn, P.L. (2007) Development of Dendrimers: Macromolecules for Use in Organic Light-Emmitting Diodes and Solar Cells. Chemical Reviews, 107, 1097-1116.

9. Zhang, M., Guo, R., Kéri, M., Bányai, I., Zheng, Y., Cao, M. and Shi, X. (2014) Impact of Dendrimer Surface Functional Groups on the Release of Doxorubicin from Dendrimer Carriers. Journal of Physical Chemistry B, 118, 1696- 1706.

10. Li, X., Haba, Y., Ochi, K., Yuba, E., Harada, A. and Kono, K. (2013) PAMAM Dendrimers with Oxyethylene Unit- Enriched Surface as Biocompatible Temperature-Sensitive Dendrimers. Bioconjugate Chemistry, 24, 282-290.

11. Fulton, D.A., Elemento, E.M., Aime, S., Chaabane, L., Botta, M. and Parker, D. (2006) Glycoconjugates of Gadolinium Complexes for MRI Applications. Chemical Communications, 10, 1064-1066.

12. Schick, I., Lorenz, S., Gehrig, D., Schilmann, A.M., Baur, H., Panthöfer, M., Fischer, K., Starnd, D., Laquai, F. and Tremel, W. (2014) Multifunctional Two-Photon Active Silica-Coated AuMnO Janus Particles for Selective Dual Functionalization and Imaging. Journal of the American Chemical Society, 136, 2473-2483.

13. Lee, C.C., MacKay, J.A., Fréchet, J.M.J. and Szoka, F.C. (2005) Designing Dendrimers for Biological Applications. Nature Biotechnology, 23, 1517-1526.

14. Bazzicalupi, C., Bianchi, A., Giorgi, C., Gratteri, P., Mariani, P. and Valtancoli, B. (2013) Metal Ion Binding by a G-2 Poly(Ethylene Imine) Dendrimer. Ion-Directed Self-Assembling of Hierarchical Mono- and Two-Dimensional Nanostructured Materials. Inorganic Chemistry, 52, 2125-2137.

15. Bergeron, R.J. (1986) Methods for the Selective Modification of Spermidine and Its Homologues. Accounts of Chemical Research, 19, 105-113.

16. Kuksa, V., Buchnan, R. and Lin, P.T. (2000) Synthesis of Polyamines, Their Derivatives, Analogues and Conjugates. Synthesis, 2000, 1189-1207.

17. Sánchez-Sandoval, A., Alvarez-Toledano, C., Gutierrez-Pérez, R. and Reyes-Ortega, Y. (2003) A Modified Procedure for the Preparation of Linear Polyamines. Synthetic Communications, 33, 481-492.

18. McMurry, J. (2001) Química Orgánica. 5th Edition, Thomson International, México.

19. Atkins, P.W., Overton, T.L., Rourke, J.P., Weller, M.T. and Armstrong, F.A. (2010) Inorganic Chemistry. Oxford University Press, New York.

20. Tanaka, K. and Toda, F. (2000) Solvent-Free Organic Synthesis. Chemical Reviews, 100, 1025-1074.

21. Nakamoto, K. (1999) Infrared and Raman Spectra of Inorganic and Coordination Compounds. John Wiley and Sons, New York.

22. Cimerman, Z., Galic, N. and Bosner, B. (1997) The Schiff Bases of Salicylaldehyde and Aminopyridines as Highly Sensitive Analytical Reagents. Analytica Chimica Acta, 343, 145-153.

23. Smith, G.S. (2003) An Investigation into the Synthesis, Characterization and Some Applications of Novel-Containing Polymers and Dendrimers of Transition Metals. Doctoral Thesis.

24. Louie, O., Massoudi, A., Vahedi, H., Asadi, H. and Sajjadifar, S. (2012) The Modification of Poly Amidoamine (PAMAM-$G_{0.5}$) by Cytosine. Engineering, 5, 103-105.

25. Radhi, M.M. and Mel-Bermani, M.F. (1990) Infrared Studies of the Conformation in Salicylaldehyde, Methylsalicylate and Ethylsalicylate. Spectrochimica Acta Part A, 46, 33-42.

26. Yildiz, M., Kiliç, Z. and Hökelek, T. (1998) Intramolecular Hydrogen Bonding and Tautomerism in Schiff Bases. Part 1. Structure of 1,8-Dif[N-2-Oxy-Phenyl-Salicylidene]-3,6-Dioxaoctane. Journal of Molecular Structure, 441, 1-10.

27. Freedman, H.H. (1961) Intramolecular H-Bonds. I. Spectroscopic Study of the Hydrogen Bond between Hydroxyl and Nitrogen. Journal of the American Chemical Society, 83, 2900-2905.

28. Colthup, N.B., Daly, L.H. and Wiberley, S.E. (1990) Introduction to Infrared and Raman Spectroscopy. 3rd Edition, Academic Press, Waltham.

29. Nazir, H., Yildiz, M., Yilmaz, H., Tahir, M.N. and Ülkü, D. (2000) Intramolecular Hydrogen Bonding and Tautomerism in Schiff Bases. Structure of N-(2-Pyridil)-2-Oxo-1-Naphthylidenemethylamine. Journal of Molecular Structure, 524, 241-250.

30. Pavia, D.L., Lampman, G.M. and Kriz, G.S. (1996) Introduction to Spectroscopy. Saunders Golden-Sunburst Series, USA.

31. Drago, R.S. (1992) Physical Methods in Chemistry. 2nd Edition, Saunders College Publishing, USA.

32. Pretsch, E., Bühlmann, P. and Affolter, C. (2000) Structure Determination of Organic Compounds. Springer, London.

CHAPTER 3

A new sorbent based on MWCNTs modification for separation/ preconcentration of trace amounts of Cd(II), Cr(III), Cu(II), Ni(II), and Pb(II) and their determination by flame atomic absorption spectrometry

Mohammad Ali Karimi1,2*, Sayed Zia Mohammadi1, Abdolhamid Hatefi-Mehrjardi2, Alireza Mohadesi1 and Javad Yarahmadi2

[1]Department of Chemistry, Payame Noor University, P.O. Box 19395-4697, Tehran, Iran

[2] Department of Chemistry & Nanoscience and Nanotechnology Research Laboratory (NNRL), Payame Noor University, P.O. Box 78185-347, Sirjan, Iran

ABSTRACT

Background

A new solid phase extraction (SPE) method has been developed for the separation/preconcentration and determination of lead, cadmium, chromium, nickel, and copper ions using a new sorbent.

Methods

The method is based on the application of multiwalled carbon nanotubes modified with ethylenediamine-*N,N*-diacetic acid (MWCNTs-EDDA) as a selective sorbent in SPE for the separation and preconcentration of Pb(II), Ni(II), Cd(II), Cu(II), and Cr(III). The determination of these heavy metals was followed by flame atomic absorption spectrometry (FAAS).

Results

All quantitative determinations were performed by FAAS, and surface modification of synthesized adsorbents has been confirmed on the basis of characterization with Fourier transform infrared spectroscopy (FTIR), scanning electron microscopy (SEM), and transmission electron microscopy (TEM). The parameters such as pH, amount of adsorbent, shaking time, eluent type and its concentration, and matrix effect have been investigated and optimized. The maximum adsorption capacity of the adsorbent at optimum conditions was found to be 51.7, 67.6, 49.9, 57.8, and 62.3 mg g^{-1} for Cd(II), Cr(III), Cu(II), Ni(II), and Pb(II), respectively.

Conclusions

In the presented study, MWCNTs were functionalized by EDDA and it was applied as an adsorbent for the separation and preconcentration of trace level of Pb, Ni, Cd, Cu, and Cr ions. This method is effective, selective, rapid, and simple for the determination of trace amounts of Pb, Ni, Cd, Cu, and Cr ions and can be successfully applied to analyze the trace amounts of mentioned ions in wastewater and herbal samples without evident interference from the complex matrix. Furthermore, the sorption has high capacity, and the precision and accuracy of the method are also satisfactory.

KEYWORDS

Solid phase extraction, Multiwalled carbon nanotubes, Ethylenediamine-*N*,*N*-diacetic acid, Heavy metals, Flame atomic absorption spectrometry

1. BACKGROUND

Heavy metal pollution on the planet is increasing day by day and has become an environmental problem throughout the world. Heavy metals, such as Cd, Cr, Cu, Ni, As, Pb, and Zn, have high solubility in the aquatic environments and can be absorbed by living organisms, enter the food chain, and accumulate in the human body. Therefore, heavy metals are considered as hazardous materials for ecosystems, animals, plants, and human beings. Heavy metals that can be found in vegetables, ground water, and wastewaters are mostly toxic even at a very low concentration, and the determination of heavy metals is one of the targets of analytical chemists (Babel and Kurniawan 2004; Barałkiewicz et al. 2010; Eskandari et al. 2006; Peker 1999; Turkoglu and Soylak 2005). However, direct determination of heavy metals is often not sensitive enough to determine heavy metal ions in environmental samples because of their low concentrations, which are near or below the detection limit of the instrument, and matrix interferences. Therefore, a preconcentration and/or separation of analytes is frequently necessary to improve the analytical detection limit, increasing the sensitivity, facilitating the calibration, enhancing the accuracy of the results, and simplifying the matrix of the sample (Baghban et al. 2012; Das et al. 2006; Hu et al. 2002a; 2002b; Palanivelu et al. 1992; Prasad et al. 2006; Rajesh and Manikandan 2008; Soylak et al. 2003).

Various separation-preconcentration techniques have been used for the separation and enrichment of trace amounts of heavy metal ions from environmental sample solutions, such as solvent (Lajunen and Kubin 1986) and cloud point (Giokas et al. 2001; Khader and Varghese 2005) extraction, ion-exchange (Möller et al. 1992), and membrane filtration (Karatepe et al. 2002). Despite many benefits, these techniques have significant disadvantages that many researchers prefer solid phase extraction (SPE) because of its notable advantages including higher preconcentration factor, rapid phase separation, short analysis time, cost saving, and low consumption of organic solvents (Absalan and Aghaei Goudi 2004; Liang et al. 2001; Pourreza and Behpour 1999). One of the main factors in SPE procedure is suitable sorbent selection because of its effect on enrichment factor and recovery (Poole 2003). Various adsorbents are used for solid phase extraction of heavy metals, for example, chitosan (Martins et al. 2005), activated carbon (de Peña et al. 1995), resins (Tuzen et al. 2004), and chromosorb 108 (Tuzen et al. 2005). Since carbon nanotubes (CNTs) were discovered by Iijima (Iijima and Ichihashi 1993), because of the large surface area, the excellent chemical and thermal stabilities, and strong adsorption ability, they have attracted great attention as SPE adsorbents for various organic pollutants and metal ions. Due to low dispersion of CNTs into solvents and limited selectivity of CNTs and oxidized CNTs for SPE, a common technique is chemical functionalization which is used relatively often to generate functional groups on the surface of CNTs (Liu et al. 2008).

In this paper, we report a new method using multiwalled carbon nanotubes modified with ethylenediamine-*N,N*-diacetic acid (MWCNTs-EDDA) as a selective sorbent in SPE for separation and preconcentration of Pb(II), Ni(II), Cd(II), Cu(II), and Cr(III) prior to their determination by flame atomic absorption spectrometry (FAAS). Characterization of synthesized adsorbent has been done using Fourier transform infrared spectroscopy (FTIR), scanning electron microscopy (SEM), and transmission electron microscopy (TEM). The effects of variables such as pH, matrix, amount of adsorbents, shaking time, eluent type, and its concentration were studied. Then, the method was applied to the analysis of wastewater and herbal samples with satisfactory results.

2. METHODS

2.1 Materials and reagents

Multiwalled carbon nanotubes (MWCNTs) of 95 % purity and a length of 1–10 μm, outer and inner diameter of 5–20 and 2–6 nm, respectively, and 3–15 number of walls were purchased from PlasmaChem GmbH (Germany, http://www.Plasmachem.com). All other reagents such as *N,N*-dicyclohexylcarbodiimide (DCC), ethyl acetate, bromo-acetic acid, ethylenediamine, sodium hydroxide, concentrated H_2SO_4 acid, concentrated HNO_3 acid, and ethanol were of at least analytical grade and obtained from Merck (Darmstadt, Germany). Standard stock solutions of Cd(II), Cr(III), Cu(II), Ni(II), and Pb(II) (1 mg mL^{-1}) were prepared by dissolving analytical or spectral purity grade chemicals $Cd(NO_3)_2.4H_2O$, $CrCl_3.6H_2O$, $Cu(NO_3)_2.3H_2O$, $Ni(NO_3)_2.6H_2O$, and $Pb(NO_3)_2$ (purchased from Merck) in Milli-Q water and further diluted daily prior to use. Standard glassware used were cleaned with HNO_3 and rinsed with Milli-Q water.

2.2 Apparatus

All ion determinations were carried out with a flame atomic absorption spectrophotometer (PG Instruments, England) with an air-acetylene flame equipped with PG instrument hallow cathode lamps. A Metrohm pH meter (model 827 pH lab, combined glass-electrode) was employed for the pH adjustment. Fourier transform infrared (FTIR) spectra (4000–400 cm^{-1}) in KBr were recorded by a FTIR Prestige-21, Shimadzu. A Philips TEM (EM 280 model) and a SEM (Tescan Vega II model) were conducted to characterize the raw MWCNTs, oxidized MWCNTs (MWCNTs-COOH), and modified MWCNTs. An ultrasonic bath (S60H Elmasonic, Germany) and a heidolph rotary vacuum evaporator were also used.

2.3 Preparation of the solid sorbent

2.3.1 Preparation of oxidized MWCNTs

For elimination of metal ions and other sorbed impurities on MWCNTs (in preparation process), it was purified with HCl (10 %) for 24 h. Then, it was filtered and treated with concentrated HNO_3 and refluxed for 3 h with stirring to introduce oxygen groups onto the MWCNT surface

(Pyrzynska 2010). The product (MWCNTs-COOH) was filtered through a 0.05-μm pore size PTFE membrane filter, washed with Milli-Q water until the pH was neutral, and dried under vacuum at 80 °C for 8 h.

2.3.2 Synthesis of MWCNTs-EDDA

The requirement ligand (EDDA) was synthesized and purified as reported in the literature (McLendon et al. 1975). The oxidized multiwalled carbon nanotubes (2.0 g) were suspended in 150 mL of water, and the synthesized EDDA (0.5 mol) was added, then 2.0 g of DCC was added into it, and the suspension was refluxed for 48 h in 100 °C. The obtained product (MWCNTs-EDDA) was vacuum filtered using a 0.05-μm pore size PTFE membrane filter, washed with excess ethanol, and dried in an oven at 80 °C for 8 h. Figure 1 shows the synthesis process of MWCNTs-EDDA schematically.

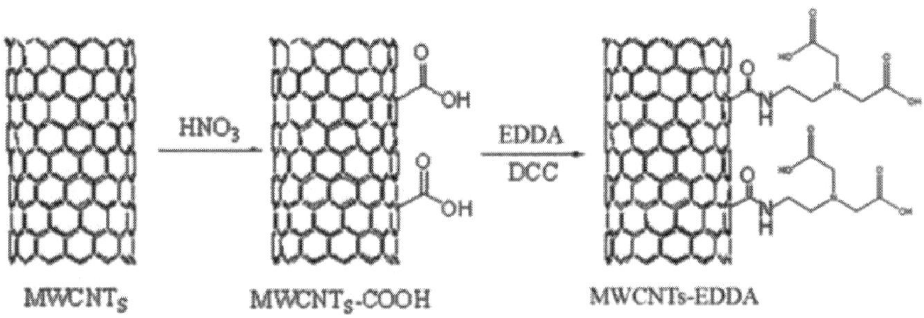

Figure 1. Schematic of MWCNTs-EDDA synthesis

2.3.3 General procedure

All adsorption experiments in this work were performed in a batch method. Water samples including four industrial wastewater and herbal samples including mushroom, black tea, and lettuce were prepared as reported in the literature (Manzoori et al. 2012). A portion of standard sample, water sample, or digested herbal sample solutions containing the studied ions were added into a 50-mL beaker, and the pH of the mixture was adjusted to the desired value with 0.1 mol L^{-1} HCl and/or 0.1 mol L^{-1} $NH_3.H_2O$, and the mixture was diluted to 10 mL with deionized water. Then, 15-mg modified MWCNTs were added to it, and the mixture was shaken for 15 min at room temperature to facilitate adsorption of the cations onto the adsorbents. After vacuum filtering, the concentrations of metal ions were determined by FAAS. Finally, 5 mL of 0.5 mol L^{-1} nitric acid was added as eluent, and the eluted ions were determined by FAAS.

3. RESULTS AND DISCUSSION

3.1 Characterization of MWCNTs-EDDA

TEM and SEM were conducted to characterize the raw MWCNTs, MWCNTs-COOH, and MWCNTs-EDDA. The representative TEM and SEM images are shown in Figs. 2 and 3, respectively. In contrast, the observed dispersing property of MWCNTs-COOH was better than that of MWCNTs, and the diameter of MWCNTs-EDDA appeared wider than that of MWCNTs and MWCNTs-COOH. These results confirm MWCNTs modified by the EDDA.

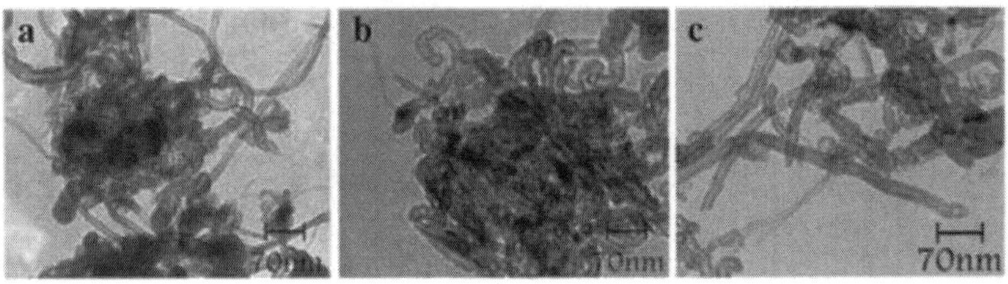

Figure 2 .TEM images of raw MWCNTs (**a**), oxidized MWCNTs (**b**), and functionalized MWCNTs by EDDA (**c**)

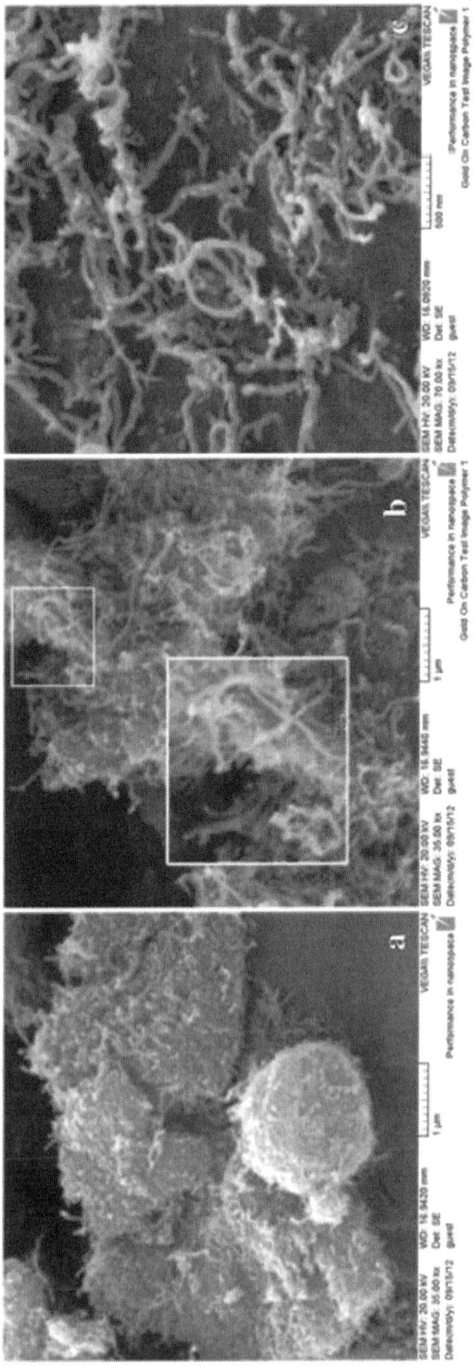

Figure 3 .SEM images of raw MWCNTs (**a**), oxidized MWCNTs (**b**), and functionalized MWCNTs by EDDA (**c**)

Formation of functionalized MWCNTs also was confirmed by IR spectroscopy. The FTIR spectrums of pure, oxidized, and functionalized MWCNTs are compared in Fig. 4. Comparison of the FTIR spectra of MWCNTs with MWCNTs-COOH shows a new bond at 1624 cm^{-1} in MWCNTs-COOH spectra which can be attributed to the stretching vibrations C=O of the carboxyl group. When MWCNTs-COOH was modified by EDDA, several new peaks appeared in the spectrum. That can be assigned as follows: the peak at 1634 cm^{-1} is due to C–O stretching vibration. The peak at 1563 cm^{-1} is caused by C–N stretching vibration and N–H bending vibration.

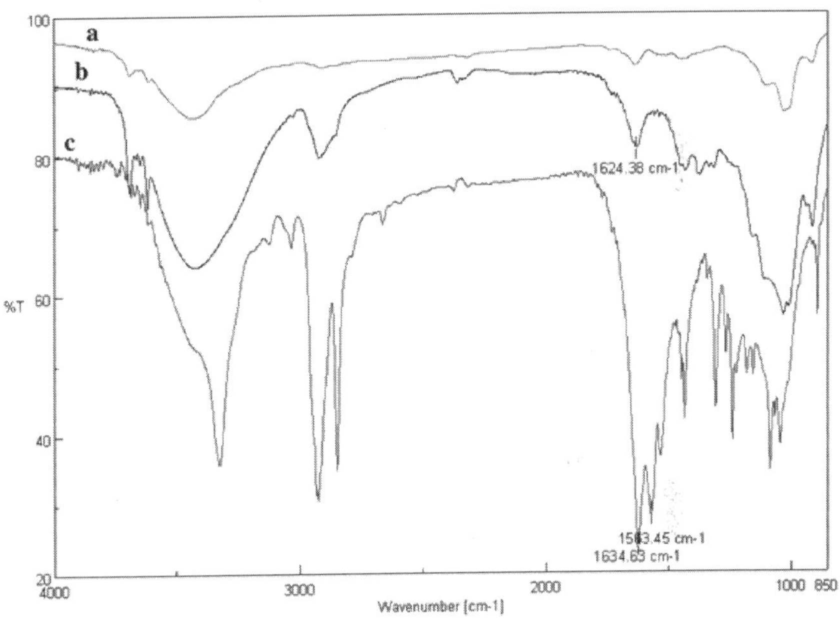

Figure 4 .FTIR spectra of raw MWCNTs (*a*), oxidized MWCNTs (*b*), and functionalized MWCNTs by EDDA (*c*)

3.2 Effect of pH on adsorption

The pH of the solution is the first parameter that should be optimized, because of two reasons: firstly, the competition of the hydrogen and metal ions for the binding site of sorbent in acidic solutions and secondly, precipitation of metal ions that complex with hydroxide (Walcarius and Delacôte 2005). The effect of pH on the adsorption of lead(II), nickel(II), cadmium(II), copper(II), and chromium(III) ions onto MWCNTs-EDDA was investigated over the range from 2 to 9 using the batch procedure. It could be seen in Fig. 5 that a quantitative extraction (>95 %) occurred in the pH range of 4.5–5.5 for Pb(II), 6.5–7.5 for Ni(II), 7.0–8.0 for Cd(II), 4.5–5.5 for Cu(II), and 3.5–5.0 for Cr(III). Therefore, pHs of 5.0, 7.0, 7.5, 5.0, and 4.0 were selected for further studies of lead, nickel, cadmium, copper, and chromium ions, respectively.

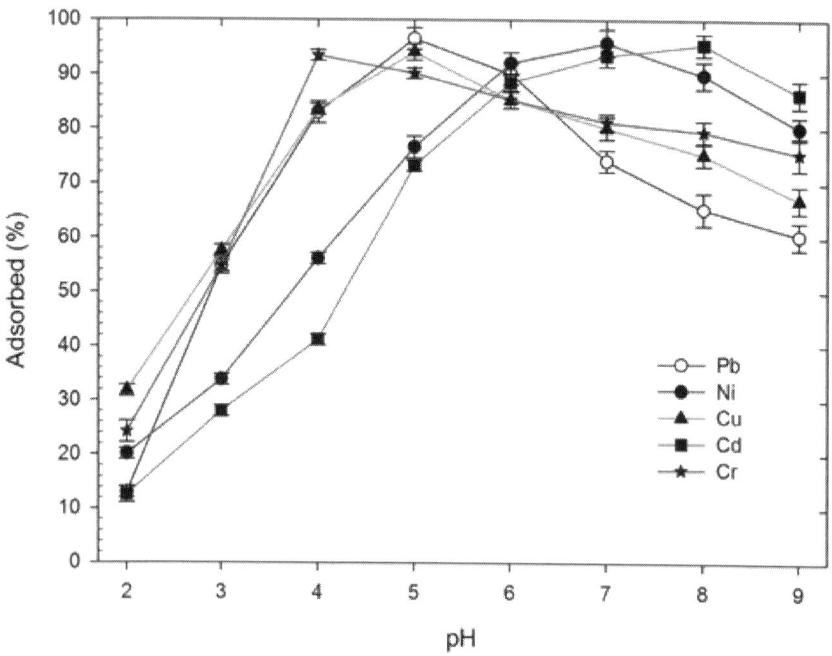

Figure 5 .Effect of pH on adsorption of heavy metal ions. Conditions: 15 mg MWCNTs-EDDA; 10 mL metallic ion solutions of Pb(II), Ni(II), and Cr(III) (10 μg mL^{-1}); 10 mL Cu(II) (5 μg mL^{-1}) and 10 mL Cd(II) (2 μg mL^{-1}) for 15 min

3.3 Effect of the sorbent amount

The effect of MWCNTs-EDDA quantity on preconcentration of considered ions was studied in the range of 1–25 mg under the optimum pH and 20-min shaking time. The results show that by increasing the amount of the sorbent, the recovery of metal ions increased due to the increasing surface area and accessible sites of the sorbent and finally reached a constant value at 15 mg of the sorbent. Therefore, 15 mg of MWCNTs-EDDA was selected for future studies.

3.4 Effect of shaking time

In solid phase extraction, the percentage extraction of metal ions always depends on the contact time of sample with the sorbent. Herein, a different shaking time (ranged from 1 to 30 min) was studied for the percentage extraction of Pb(II), Ni(II), Cr(III) (10 mL, 10 $\mu g\ mL^{-1}$), Cu(II) (10 mL, 5 $\mu g\ mL^{-1}$), and Cd(II) (10 mL, 2 $\mu g\ mL^{-1}$), with 15 mg of the sorbent. The results show that the adsorption of the mentioned ions at first 10 min was higher than 95 %, but 15 min was used for future studies.

3.5 Effect of eluent type and concentration

As shown in Fig. 5, the adsorption of metal ions decreases at low pHs, so a series of acidic solutions including HNO_3, HCl, and H_2SO_4 were used for elution of the retained Cd(II), Cr(III), Cu(II), Ni(II), and Pb(II). The obtained results (Table 1) showed that 5 mL of 0.5 $mol\ L^{-1}$ HNO_3 solution was suitable for complete elution of retained heavy metals.

3.6 Effect of sample volume

Due to low concentration of heavy metals in natural samples, it is necessary to obtain the maximum volume in the SPE. For this purpose, various sample volumes in the range of 10–850 mL containing 100 μg of

Pb(II), Ni(II), and Cr(III), 50 μg of Cu(II), and 20 μg of Cd(II) were treated according to the general procedure. It was found that quantitative recoveries (>95 %) were obtained for the sample volumes up to 650 mL for Pb(II), Ni(II), Cu(II), and Cd(II) and 600 mL for Cr(III) ions (Fig. 6). The final solution volume was 5.0 mL, and therefore, the preconcentration factors were 120 for Cr(III) and 130 for Pb(II), Ni(II), Cu(II), and Cd(II) ions.

Table 1. Effect of different eluents on the recoveries of heavy metal ions

Eluent	Concentration (mol L^{-1})	Recovery (%)				
		Pb(II)	Ni(II)	Cr(III)	Cu(II)	Cd(II)
HNO_3	0.1	93.3	88.5	88.7	89.9	92.5
	0.3	96.7	95.7	94.1	95.7	97.8
	0.5	98.4	99.6	97.7	98.1	99.7
H_2SO_4	0.1	78.5	83.4	86.7	92.4	87.5
	0.3	80.3	87.1	88.9	96.5	89.4
	0.5	84.2	89.9	91.2	98.7	91.4
HCl	0.1	84.6	80.2	78.6	83.5	88.4
	0.3	86.7	82.4	81.1	88.8	91.2
	0.5	88.2	86.9	84.0	91.2	96.3

3.7 Maximum adsorption capacity

The adsorption capacity of MWCNTs-EDDA investigated using the batch method as reported by Maquieira et al. (1994). A portion of sample solutions of various concentrations (10–100 μg mL^{-1}, 10 mL) of Cd(II), Cr(III), Cu(II), Ni(II), and Pb(II) were adjusted to the appropriate pH, and all processes of separation and preconcentration were applied on it. The maximum adsorption capacities of MWCNTs-EDDA for Cd(II), Cr(III), Cu(II), Ni(II), and Pb(II) were found to be 51.7, 62.6, 49.9, 57.8, and 62.3 mg g^{-1}, respectively. Our results show that untreated and oxidized MWCNTs have much lower adsorption than new synthesized adsorbent.

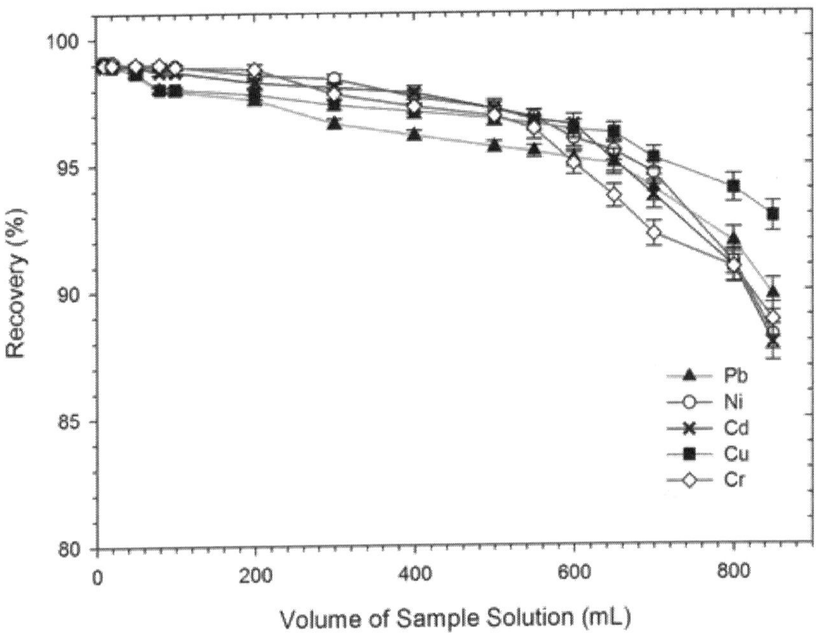

Figure 6 .ffect of sample volume on adsorption of heavy metal ions. Conditions: The conditions are same as Fig. 5 except pH 5.0, 6.0, 9.0, 8.0 and 7.0 for Pb(II), Ni(II), Cd(II), Cu(II) and Cr(III) ions, respectively

3.8 Adsorption isotherms

Adsorption isotherms were obtained at 25 °C and pH 5.0, 7.0, 7.5, 5.0, and 4.0 for Pb(II), Ni(II), Cd(II), Cu(II), and Cr(III) ions, respectively, and by varying the initial concentration of these ions (Fig. 7a, b). Langmuir and Freundlich isotherms are two commonly used statistical isotherm models. The Langmuir isotherm demonstrates a monolayer coverage of adsorbate with homogeneous sorption energies, while Freundlich isotherm demonstrates a non-uniform distribution of heat of adsorption over the surface with the possibility of multilayer adsorption

(Yu et al. 2007). The adsorption data were correlated with the isotherm models of Langmuir (Eq. 1) and Freundlich (Eq. 2)

$$q = \frac{(bq_{\max}C)}{(1+bC)} \text{ or } \frac{C}{q} = \frac{1}{(bq_{\max})} + \frac{C}{q_{\max}} \tag{1}$$

$$q = k_f C^n \text{ or } \log q = \log k_f + n \log C \tag{2}$$

where C is the equilibrium concentration of metal ions (μg mL^{-1}), q is the equilibrium amount of metal ion adsorbed per unit weight of adsorbents (mg g^{-1}), q_{max} is the maximum adsorption capacity of the adsorbents for metal ions (mg g^{-1}), b is the Langmuir constant and is a measure of the energy of adsorption (L mg^{-1}), k_f is the Freundlich constant representing the adsorption capacity, and n is a constant depicting the adsorption intensity. The Langmuir and Freundlich constants were obtained by fitting the adsorption equilibrium data to the isotherm models; they are listed in Table 2. It can be noticed that the R values for the Langmuir model are higher than those of the Freundlich model, thus indicating that the Langmuir model better describes adsorption onto new sorbent.

3.9 Effect of interfering ions

Always in real samples, there are many ions and determination of one of them may be influenced by others. Therefore, the effects of interfering ions on the determination of Cd(II), Cr(III), Cu(II), Ni(II), and Pb(II) were examined using 10 μg mL^{-1} of Pb, Ni, and Cr, 5 μg mL^{-1} of Cu, and 2 μg mL^{-1} of Cd containing the added interfering ions. The tolerance limit of the interfering ions, defined as the maximum concentration of interfering ion causing a relative error of adsorbed Cd(II), Cr(III), Cu(II), Ni(II), and Pb(II), is less than ±5 %. The results listed in Table 3 show that the suggested method had a good selectivity and can be applied in the determination of studied ions in real samples without any interference of other ions.

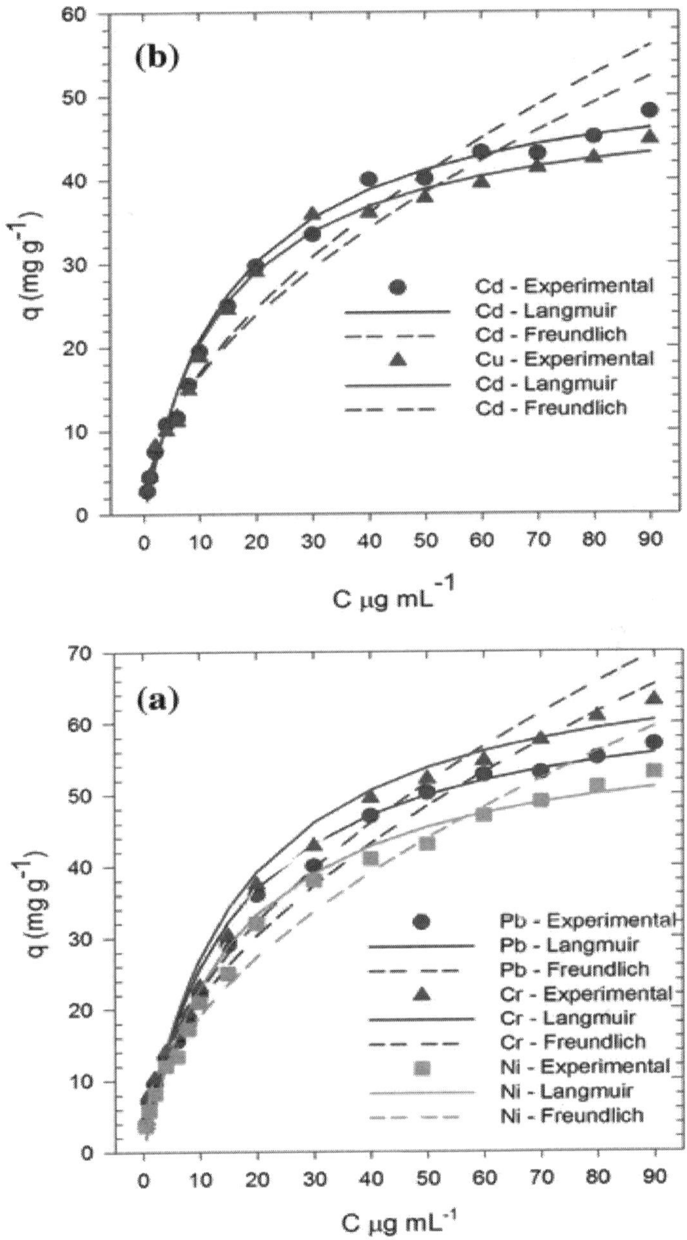

Figure 7 .Adsorption isotherms of **a** Pb(II), Cr(III), and Ni(II) and **b** Cd(II) and Cu(II) by MWCNTs-EDDA at initial metal ions concentration from 0.5 to 90 µg mL^{-1}, pH 5.0, 6.0, 9.0, 8.0, and 7.0 for Pb(II), Ni(II), Cd(II), Cu(II), and Cr(III) ions, respectively, and temperature of 25 ± 0.2 °C for 24 h. The data are fitted by Langmuir model and Freundlich model

Table 2. Langmuir and Freundlich constants for the adsorption of Pb(II), Cu(II), Cr(III), Cd(II), and Ni(II) by MWCNTs-EDDA

Metal ions	Langmuir			Freundlich		
	q_{max} (mg g^{-1})	b	R	k_f	n	R
Pb(II)	65.351	0.06468	0.9867	6.5433	0.5114	0.9712
Cr(III)	71.45	0.06063	0.9853	6.9347	0.5132	0.9825
Ni(II)	60.19	0.06155	0.9879	5.7904	0.5177	0.9786
Cd(II)	54.26	0.06267	0.9907	4.8323	0.5445	0.9597
Cu(II)	50.12	0.06905	0.9875	4.9930	0.5217	0.9503

Table 3. Effect of interferences ions on preconcentration and determination of heavy metal ions

Interfering ion					Tolerance limit (mg L^{-1})
Pb	Cu	Ni	Cd	Cr	
Ca^{2+}, Mg^{2+}, Na^{+}, K^{+}, Ni^{2+}	Na^{+}, K^{+}, Mn^{2+}, Ca^{2+}	Na^{+}, K^{+}, Ca^{2+}	K^{+}, Na^{+}, Fe^{3+}	K^{+}, Na^{+}, NH_4^{+}	650
Ni^{2+}, Mn^{2+}, Cu^{2+}, Fe^{3+}	Pb^{2+}, NO_3^{-}, Mg^{2+}, NH_4^{+}	Mn^{2+}, Cd^{2+}, Pb^{2+}, Fe^{3+}	Mn^{2+}, NO_3^{-}, NH_4^{+}, Mg^{2+}	Mn^{2+}, Ni^{2+}, Cu^{2+}, NO_3^{-},	550
Cr^{3+}, Zn^{2+}, Cd^{2+}	Ni^{2+}, Hg^{2+}, Ba^{2+}, Cr^{3+}, Zn^{2+}, CH_3COO^{-}	Mg^{2+}, Cu^{2+}, NO_3^{-}, NH_4^{+}	Ca^{2+}, Pb^{2+}, Zn^{2+}, Hg^{2+}, Cu^{2+}	Ca^{2+}, Mg^{2+}, Fe^{3+}, Zn^{2+}, Hg^{2+}	400
Co^{2+}, NH_4^{+}, Ba^{2+}	$C_2O_4^{2-}$, SO_4^{2-}, Co^{2+}, Fe^{3+}	Cr^{3+}, Zn^{2+}, Hg^{2+}, Ba^{2+}, Co^{2+}	Ni^{2+}, Co^{2+}, Br^{-}	Pb^{2+}, Co^{2+}, Ba^{2+}, $C_2O_4^{2-}$, CH_3COO^{-}	250
Hg^{2+}, NO^{3-}, CH_3COO^{-}	Br^{-}	$C_2O_4^{2+}$, Br^{-}	$C_2O_4^{2-}$, Ba^{2+}	SO_4^{2-}, Br^{-}	150
SO_4^{2-}, Br^{-}, $C_2O_4^{2-}$		CH_3COO^{-}, SO_4^{2-}	SO_4^{2-}, CH_3COO^{-}		100

3.10 Analytical performance and method validation

In order to show the validation of the proposed method, the analytical features of the method such as limit of detection (LOD), linear range of the calibration curve, and precision of ions were examined. Under the optimal experimental conditions, eight standard solutions of Cd(II), Cr(III), Cu(II), Ni(II), and Pb(II) were analyzed by the procedure. The LODs were found to be 0.04, 0.15, 0.14, 0.21, and 0.89 μg mL^{-1} for Cd, Cr, Cu, Ni, and Pb ions, respectively. The linear ranges of calibration curves for Cr(III), Ni(II), and Pb(II) were 0.5–10.0 μg mL^{-1}, 0.5–5.0 μg mL^{-1} for Cu(II), and 0.2–2.0 μg mL^{-1} for Cd(II) with correlation coefficients of 0.991, 0.996, 0.997, 0.999, and 0.998, respectively. The regression equations for the lines were $A = 0.014C_{Pb} + 0.001$, $A = 0.059C_{Ni} + 0.034$, $A = 0.008C_{Cr} + 0.010$, $A = 0.261C_{Cd} + 0.006$, and $A = 0.085C_{Cu} + 0.001$, where A is the absorbance and C_{Pb}, C_{Ni}, C_{Cr}, C_{Cd}, and C_{Cu} are the concentrations of Pb(II), Ni(II), Cr(III), Cd(II), and Cu(II) in μg mL^{-1}, respectively. Intra-day and inter-day relative standard deviations for three individual experiments and 3 days of analysis in water solution were 1.82–3.31 % and 1.63–3.57 %, respectively (Table 4).

Table 4. Intra-day and inter-day relative standard deviations for three individual experiments and 3 days of analysis in water solution

Days of analysis	Inter-day RSD (%)				
	Pb	Ni	Cd	Cu	Cr
Day 1	3.16	2.19	2.23	1.8	3.1
Day 2	2.83	2.94	2.13	1.88	2.79
Day 3	3.57	1.63	1.87	2.14	2.3
Intra-day RSD (%)	3.12	2.17	2.01	1.82	3.31

The RSDs for three replicate measurements of 10.0 μg mL^{-1} Cr, Ni, and Pb ions, 2.0 μg mL^{-1} Cd ions, and 5.0 μg mL^{-1} Cu ions

The RSDs for three replicate measurements of 10.0 $\mu g\ mL^{-1}$ Cr, Ni, and Pb ions, 2.0 $\mu g\ mL^{-1}$ Cd ions, and 5.0 $\mu g\ mL^{-1}$ Cu ions

3.11 Analytical applications

In order to establish the validity of the proposed method, it has been applied for the determination of trace amounts of Pb(II), Ni(II), Cd(II), Cu(II), and Cr(III) in wastewaters from Sarcheshmeh Copper factory and Yazd sewage treatment plant and herbal samples (black tea, mushroom, and lettuce). The results are listed in Tables 5 and 6. According to these data, the added ions can be quantitatively recovered from the samples by the proposed procedure. These results indicate that the method is a selective SPE and determination procedure of trace Pb(II), Ni(II), Cd(II), Cu(II), and Cr(III) in environmental samples.

4. CONCLUSIONS

In the presented study, MWCNTs were functionalized by EDDA and it was applied as an adsorbent for the separation and preconcentration of the trace level of Pb, Ni, Cd, Cu, and Cr ions. This method is effective, selective, rapid, and simple for the determination of the trace amounts of Pb, Ni, Cd, Cu, and Cr ions and can be successfully applied to analyze the trace amounts of the mentioned ions in wastewater and herbal samples without evident interference from the complex matrix. Comparison of this method and some of the previously reported preconcentration methods (Table 7) shows that the LOD and preconcentration factor of this method is comparable or better than achieved by other methods described in the literatures. Furthermore, the sorption has high capacity, and the precision and accuracy of the method are also satisfactory.

Table 5 Recovery results of heavy metal ions in wastewater samples

Metallic ion	Wastewater (Copper factory, Sarcheshmeh, Rafsanjan)			Waste water (sewage treatment pla	
	Added	Found	Recovery (%)	Added	Found
Pb(II) (μg mL^{-1})[a]	0	2.20 (±0.14)	–	0	0.46 (±0.10)
	5.0	7.11 (±0.21)	98.2	5.0	5.24 (±0.16)
	10.0	12.24 (±0.17)	100.4	10.0	10.31 (±0.08)
Ni(II) (μg mL^{-1})[a]	0	1.10 (±0.11)	–	0	0.29 (±0.10)
	5.0	6.12 (±0.15)	100.4	5.0	5.37 (±0.12)
	10.0	11.08 (±0.12)	99.8	10.0	10.36 (±0.14)
Cd(II) (μg mL^{-1})[a]	0	0.50 (±0.09)	–	0	0.42 (±0.13)
	5.0	5.44 (±0.14)	98.8	5.0	5.41 (±0.16)
	10.0	10.52 (±0.19)	100.2	10.0	10.47 (±0.11)
Cu(II) (μg mL^{-1})[a]	0	6.21 (±0.13)	–	0	0.59 (0.10)
	5.0	11.13 (±0.15)	98.4	5.0	5.52 (±0.17)
	10.0	16.10 (±0.18)	98.9	10.0	10.64 (±0.15)
Cr(III) (μg mL^{-1})[a]	0	0.57 (±0.10)	–	0	0.56 (±0.17)
	5.0	5.60 (±0.15)	100.6	5.0	5.50 (±0.18)
	10.0	10.47 (±0.15)	99.0	10.0	10.60 (±0.13)

[a]Mean ± standard deviation ($n = 5$)

Table 6 Recovery results of heavy metal ions in herbal samples

Sample	Cu(II)[a]			Cr(III)[a]		
	Added[b]	Found[b]	Recovery (%)	Added[b]	Found[b]	Recovery (%)
Lettuce	–	–	–	0	0.43 ± 0.08	–
	–	–	–	5.0	5.4 ± 0.11	99.4
	–	–	–	10.0	10.37 ± 0.17	99.4
Mushroom	0	13.71 ± 0.59	–	–	–	–
	5.0	18.58 ± 0.14	97.4	–	–	–
	10.0	23.78 ± 0.23	100.7	–	–	–
Black tea	0	19.68 ± 1.09	–	0	5.39 ± 1.12	–
	5.0	24.46 ± 0.41	95.6	5.0	10.24 ± 1.05	97.0
	10.0	29.75 ± 0.39	103.9	10.0	15.14 ± 0.82	97.5

[a]Mean ± standard deviation ($n = 5$)
[b]Unites by mg kg^{-1}

Table 7. Comparison of the proposed method with some recent studies using multiwalled carbon nanotubes as solid phase extractor

Analytes	Adsorbent	Detection method	Enrichment factor	RSD (%)	Detection limit (ng mL^{-1})	Reference
Pb	MWCNTs	FAAS	20	<2.5	4.9, 8.0	(Ozcan et al. 2010)
Pb, Cu, Cd, Ni	APDC-MWCNTs	FAAS	80	<5	0.3–0.6	(Tuzen et al. 2008)
Pb, Cd	MnO_2-CNTs	ETAAS	100	<3.2	4.4, 1.5	(Yang et al. 2011)
Ni	Activated carbon	FAAS	100	2.28	1.72	(Kiran et al. 2007)
Cu, Ni	MWCNT-hexahistidine -tagged protein	FAAS	29.28	2.4	0.3, 0.63	(Liu et al. 2009)
Cr	MWCNT	FAAS	100	<9	0.9	(Tuzen and Soylak 2007)
Cu, Ni	MWCNTs impregnated with D2EHPA	FAAS	25	<10	50	(Vellaichamy and Palanivelu 2011)
Cu, Pb, Ni	MWCNTs	FAAS	100	2.4–3.4	1.0–2.6	(Ghaedi et al. 2012)
Pb				3.16	0.89	
Ni				2.19	0.21	
Cr	MWCNTs-EDDA	FAAS	130	3.34	0.15	This work
Cd				2.23	0.04	
Cu				1.89	0.14	

ACKNOWLEDGEMENTS

The Iran National Science Foundation (INSF) is gratefully acknowledged for financial support of this work (Project No. 90002933).

COMPETING INTERESTS

The authors declare that they have no competing interests.

AUTHORS' CONTRIBUTIONS

MAK, AM and SZM conceived of the original idea and design of the work. JY performed the collection of water and herbal samples, preparation of reagents and solutions and execution of experiments. MAK and AHM performed the data interpretation and manuscript writing. All authors read and approved the final manuscript.

REFERENCES

1. Absalan G, Aghaei Goudi A (2004) Optimizing the immobilized dithizone on surfactant-coated alumina as new sorbent for determination of silver. Sep Purif Technol 38:209–214
2. Babel S, Kurniawan TA (2004) Cr(VI) removal from synthetic wastewater using coconut shell charcoal and commercial activated carbon modified with oxidizing agents and/or chitosan. Chemosphere 54:951–967
3. Baghban N, Haji Shabani AM, Dadfarnia S (2012) Solid phase extraction and flame atomic absorption spectrometric determination of trace amounts of cadmium and lead in water and biological

samples using modified TiO_2 nanoparticles. Int J Environ Anal Chem 93:1367–1380

4. Barałkiewicz D, Hanć A, Gramowska H (2010) Simultaneous determination of Cd, Cr, Cu, Ni, Pb and Zn in sewage sludge by slurry introduction ICP-OES method. Int J Environ Anal Chem 90:1025–1035
5. Das AK, Chakraborty R, Cervera ML, de la Guardia M (2006) Analytical techniques for the determination of bismuth in solid environmental samples. Trends Anal Chem 25:599–608
6. de Peña YP, Gallego M, Valcárcel M (1995) On-line sorbent extraction, preconcentration and determination of lead by atomic absorption spectrometry. Talanta 42:211–218
7. Eskandari H, Saghseloo AG, Chamjangali MA (2006) First- and second-derivative spectrophotometry for simultaneous determination of copper and cobalt by 1-(2-pyridylazo)-2-naphthol in Tween 80 micellar solutions. Turk J Chem 30:49–63
8. Ghaedi M, Montazerozohori M, Biyareh MN, Mortazavi K, Soylak M (2012) Chemically bonded multiwalled carbon nanotubes as efficient material for solid phase extraction of some metal ions in food samples. Int J Environ Anal Chem 93:528–542
9. Giokas DL, Paleologos EK, Tzouwara-Karayann SM, Karayannis MI (2001) Single-sample cloud point determination of iron, cobalt and nickel by flow injection analysis flame atomic absorption spectrometry-application to real samples and certified reference materials. J Anal Atom Spectrom 16:521–526
10. Hu Q, Guangyu Y, Huang Z, Yin J (2002a) Spectrophotometric determination of silver with 2-(2-quinolylazo)-5-diethylaminoaniline. Talanta 58:467–473
11. Hu Q, Yang G, Yin J, Yao Y (2002b) Determination of trace lead, cadmium and mercury by on-line column enrichment followed by RP-HPLC as metal-tetra-(4-bromophenyl)-porphyrin chelates. Talanta 57:751–756
12. Iijima S, Ichihashi T (1993) Single-shell carbon nanotubes of 1-nm diameter. Nature 363:603–605

13. Karatepe AU, Soylak M, Elci L (2002) Separation/preconcentration of Cu(II), Fe(III), Pb(II), Co(II), and Cr(III) in aqueous samples on cellulose nitrate membrane filter and their determination by atomic absorption spectrometry. Anal Lett 35:1561–1574

14. Khader AMA, Varghese A (2005) Simultaneous determination of titanium and molybdenum in alloy steels using derivative spectrophotometry in neutral micellar medium. Indian J Chem Technol 12:707–707

15. Kiran K, Suresh Kumar K, Suvardhan K, Janardhanam K, Chiranjeevi P (2007) Retracted: Preconcentration and solid phase extraction method for the determination of Co, Cu, Ni, Zn and Cd in environmental and biological samples using activated carbon by FAAS. J Hazard Mater 147:15–20

16. Lajunen LHJ, Kubin A (1986) Determination of trace amounts of molybdenum in plant tissue by solvent extraction-atomic-absorption and direct-current plasma emission spectrometry. Talanta 33:265–270

17. Liang P, QinY HB, Peng T, Jiang Z (2001) Nanometer-size titanium dioxide microcolumn on-line preconcentration of trace metals and their determination by inductively coupled plasma atomic emission spectrometry in water. Anal Chim Acta 440:207–213

18. Liu Y, Li Y, Yan XP (2008) Preparation, characterization, and application of l-cysteine functionalized multiwalled carbon nanotubes as a selective sorbent for separation and preconcentration of heavy metals. Adv Funct Mater 18:1536–1543

19. Liu Y, Li Y, Wu ZQ, Yan XP (2009) Fabrication and characterization of hexahistidine-tagged protein functionalized multi-walled carbon nanotubes for selective solid-phase extraction of Cu^{2+} and Ni^{2+}. Talanta 79:1464–1471

20. Manzoori J, Amjadi M, Darvishnejad M (2012) Separation and preconcentration of trace quantities of copper ion using modified alumina nanoparticles, and its determination by flame atomic absorption spectrometry. Microchim Acta 176:437–443

21. Maquieira A, Elmahadi HAM, Puchades R (1994) Immobilized cyanobacteria for online trace metal enrichment by flow injection atomic absorption spectrometry. Anal Chem 66:3632–3638

22. Martins AO, da Silva EL, Laranjeira MCM, de Fávere VT (2005) Application of chitosan functionalized with 8-hydroxyquinoline: determination of lead by flow injection flame atomic absorption spectrometry. Microchim Acta 150:27–33

23. McLendon G, Motekaitis RJ, Martell AE (1975) Cobalt complexes of ethylenediamine-n, n'-diacetic acid and ethylenediamine-n, n-diacetic acid. Two-nitrogen oxygen carriers. Inorg Chem 14:1993–1996

24. Möller P, Dulski P, Luck J (1992) Determination of rare earth elements in seawater by inductively coupled plasma-mass spectrometry. Spectrochim Acta B 47:1379–1387

25. Ozcan SG, Satiroglu N, Soylak M (2010) Column solid phase extraction of iron(III), copper(II), manganese(II) and lead(II) ions food and water samples on multi-walled carbon nanotubes. Food Chem Toxicol 48:2401–2406

26. Palanivelu K, Balasubramanian N, Ramakrishna TV (1992) A chemical enhancement method for the spectrophotometric determination of trace amounts of arsenic. Talanta 39:555–561

27. Peker I (1999) The solubilities of Zn, Sb and Se found in anode slime. Asian J Chem 11:979–986

28. Poole CF (2003) New trends in solid-phase extraction. Trends Anal Chem 22:362–373

29. Pourreza N, Behpour M (1999) Column preconcentration of aluminum using eriochrome cyanine R and methyltrioctylammonium chloride adsorbent supported on naphthalene with subsequent spectrophotometric determination. Microchem J 63:250–256

30. Prasad K, Gopikrishna P, Kala R, Rao TP, Naidu GRK (2006) Solid phase extraction vis-à-vis coprecipitation preconcentration of cadmium and lead from soils onto 5,7-dibromoquinoline-8-ol embedded benzophenone and determination by FAAS. Talanta 69:938–945

31. Pyrzynska K (2010) Carbon nanostructures for separation, preconcentration and speciation of metal ions. Trends Anal Chem 29:718–727

32. Rajesh N, Manikandan S (2008) Spectrophotometric determination of lead after preconcentration of its diphenylthiocarbazone complex on a Aamberlite XAD-1180 column. Spectrochim Acta A 70:754–757

33. Soylak M, Karatepe AU, ElÇİ L, Doğan M (2003) Column preconcentration/separation and atomic absorption spectrometric determinations of some heavy metals in table salt samples using amberlite XAD-1180. Turk J Chem 27:235–242

34. Turkoglu O, Soylak M (2005) Spectrophotometric determination of copper in natural waters and pharmaceutical samples with chloro(phenyl) glyoxime. J Chin Chem Soc 52:575–579

35. Tuzen M, Soylak M (2007) Multiwalled carbon nanotubes for speciation of chromium in environmental samples. J Hazard Mater 147:219–225

36. Tuzen M, Soylak M, ElÇİ L, Doğan M (2004) Column solid phase extraction of copper, iron, and zinc ions at trace levels in environmental samples on Amberlite XAD-7 for their flame atomic absorption spectrometric determinations. Anal Lett 37:1185–1201

37. Tuzen M, Soylak M, Elci L (2005) Multi-element pre-concentration of heavy metal ions by solid phase extraction on chromosor B 108. Anal Chim Acta 548:101–108

38. Tuzen M, Saygi KO, Soylak M (2008) Solid phase extraction of heavy metal ions in environmental samples on multiwalled carbon nanotubes. J Hazard Mater 152:632–639

39. Vellaichamy S, Palanivelu K (2011) Preconcentration and separation of copper, nickel and zinc in aqueous samples by flame atomic absorption spectrometry after column solid-phase extraction onto MCWNTs impregnated with D2EHPA-TOPO mixture. J Hazard Mater 185:1131–1139

40. Walcarius A, Delacôte C (2005) Mercury(ii) binding to thiol-functionalized mesoporous silicas: critical effect of ph and sorbent properties on capacity and selectivity. Anal Chim Acta 547:3–13

41. Yang B, Gong Q, Zhao L, Sun H, Ren N, Qin J, Xu J, Yang H (2011) Preconcentration and determination of lead and cadmium in water samples with a MnO_2 coated carbon nanotubes by using ETAAS. Desalination 278:65–69

42. Yu J, Tong M, Sun X, Li B (2007) Cystine-modified biomass for Cd(II) and Pb(II) biosorption. J Hazard Mater 143:277–284

CHAPTER 4

Use of Flame Atomic Absorption Spectroscopy and Multivariate Analysis for the Determination of Trace Elements in Human Scalp

Sayo O. Fakayode*, **Sri Lanka Owen, David A. Pollard, Mamudu Yakubu**

Department of Chemistry, Winston-Salem State University, Winston-Salem, USA

ABSTRACT

The analysis of trace elements in human hair for use as biomarkers continues to generate considerable interest in environmental and bioanalytical studies, medical diagnostics, and forensic science. This study investigated the concentrations of essential and toxic elements (Fe, Mg, Ca, Cu, Zn, Cr, Cd, and Pb) using flame atomic absorption spectroscopy (FAAS) in human scalp hair obtained from subjects living in Forsyth County, North Carolina, USA. The influence of age, sex, race, and smoking habits on the levels of trace elements in the hair samples were also investigated. Additionally, analyses were subjected to a statistical, regression, and principal component analysis to evaluate inter-elemental association and possible pattern recognition in hair samples. Furthermore, Ca/Mg and Zn/Cu ratios, which are often used to evaluate the degree of Ca and Cu utilization in humans and as markers for various health related issues including, atherosclerosis, hypertension, insulin sensitivity, and pancreatic cancer, were calculated. The overall mean concentrations of Fe (25 μg/g), Ca (710 μg/g), Mg (120 μg/g), Zn (190 μg/g), Cu (12 μg/g), and Cr (0.20 μg/g) were found in hair samples. The trace element concentrations varied widely in hair samples as demonstrated by large range of concentrations obtained for each element. However, levels of Cd and Pb elements of <0.030 μg/g were detected in hair sample. In general, the levels of the trace elements in hair samples were poorly correlated. However, significant correlations were found between Ca and Mg ($r = 0.840$, $p = 0.05$). The levels of Fe, Ca, Mg, Zn, Cu, and Cr in hair samples and the calculated Ca/Mg and Zn/Cu ratios were found to be largely correlated with age, race, sex, and smoking habits.

KEYWORDS

Human Scalp Hair; Biomonitoring; Trace Element Concentration; Flame Atomic Absorption Spectroscopy; Regression and Principal Component Analysis; Inter-Element Association

1. INTRODUCTION

Trace elements, including Fe, Ca, Mg, Zn, and Cu, play critical roles in proper human body development and metabolic activity [1-3]. For example, Fe is required in the diet for proper functioning of the liver and hemoglobin, a protein in humans primarily responsible for the transportation and distribution of oxygen from the lungs to various human organs [3]. Iron is also a critical component of myoglobin, an oxygen storing heme protein residing in the cell and is responsible for the color of meat [4-7]. Iron deficiency may result in insomnia, stunted growth, decreased immune function and inhibition of hemoglobin synthesis, particularly in women and children, resulting in anemia [8,9]. Calcium and magnesium are necessary for effective teeth and bone structure development, facilitating transmission of nerves impulses, carbohydrate and protein metabolism, activation of various enzymatic reactions, and enhancement of the absorption of phosphorus and vitamins [3]. Zinc and copper promote normal body metabolism, are critical in normal genetic expression, and are essential co-enzymes, catalyzing various enzymatic reactions [1,10]. These trace elements are required at certain concentrations which can be primarily obtained from sufficiently balanced diet and food supplements.

In contrast, heavy metals such as Pb and Cd are non essential elements and their presence in the human body can be harmful, with serious health consequences. For instance, health hazards, including reduction of child intelligence quotients, hypertension, depression, mental disorders, cancer diseases, and deoxyribonucleic acid damage have been associated with heavy metals at high concentrations [11-15]. Humans are often exposed to heavy metals through various routes, including the consumption of heavy metal contaminated foods or water, occupational exposure in working environments, lead paints, contaminated toys, industrial activity, vehicular emissions, and environmental exposure [16-30]. The critical role of essential elements in normal metabolic activity and the potential negative health consequences of toxic heavy metals necessitate the need for routine biomonitoring of the levels of trace metals elements.

Toward these efforts, various studies have utilized a wide range of human specimens including, human serum, blood, tissues, urine, and saliva as biomarkers for effective biomonitoring of trace element concentrations [31- 35]. However, these biomarkers have some major drawbacks including, sample susceptibility to contamination, sample decomposition, and instability. Additionally, some human specimens may involve invasive sample collection. Human scalp hair has lately become more preferable to the analysis of traditional human specimens because many trace elements are more concentrated with longer residence times in hair compared to other human specimens. Besides, hair samples are easy to collect, noninvasive, easy to store, and relatively inexpensive. Thus, human hair samples have been effectively used as biomarkers for the determination of trace element concentrations in environmental and bioanalytical studies, clinical and medical diagnosis, occupational health, recreational and therapeutic use, and forensic science [36-56].

We hypothesized that the combined use of analytical spectroscopy, multivariate analysis, statistics, and human scalp hair can be utilized to map the level of essential and potentially toxic trace elements in human subjects. We also hypothesized that race would possibly influence the level of trace elements in the scalp hair of human subjects. Consequently, this study reports for the first time, a comprehensive and combined use of analytical spectroscopy, statistics, multivariate analysis, and human scalp hair as a biomarker for assessing trace element (Fe, Mg, Ca, Cd, Pb, Cu, Zn, and Cr) concentrations in human subjects. The study also reports for the first time, the influence of race on the levels of trace element concentrations and the use of multivariate analysis for trace element pattern recognition in scalp hair of human subjects. Furthermore, the study reports for the first time the influence of race, age, race, sex, on the Ca/Mg and Zn/Cu ratios (which are often used to evaluate the degree of Ca and Cu utilization in human and markers for various health related issues including, atherosclerosis, hypertension, insulin sensitivity, and pancreatic cancer) in the scalp hair of human subjects.

2. MATERIALS AND METHODS

2.1. Sample Collection and Sample Preparation

Human scalp hair samples were collected in three cities (Winston-Salem, Clemmons, and Kernersville) in Forsyth County, North Carolina, USA. Forsyth County is 409.60 square miles in area, with a population of 350,670 [57]. Females and males account for 52% and 48%, respectively of the population. In addition, people under 18 years of age account for 24.6% of the sample population while those older than 18 years are make-up of 73.4% of the population in Forsyth County. The demographic racial composition of Forsyth County consists of Caucasians (majority 62.3%) and non-Caucasians (minority, 37.7%). Non-Caucasians were made up of African Americans, 26%; Hispanics,

11.9%; and Asians, 1.9%. Scalp hair samples were collected from 125 participants using stainless steel scissors during regular haircut periods at various barber shops and beauty salons over a period of three weeks in June 2011. The sample population consists of 60 males and 65 females aged between 5 and 70 years old. In addition, the hair samples were collected from different races and demographics consisting of Caucasian (n = 87) and non-Caucasians (n = 38). Non-Caucasians were made up of African Americans, Hispanics, and Asians. Only eighteen of the people from whom hair samples were collected declared themselves as tobacco product smokers while 107 identified themselves as nonsmokers.

The collected hair samples were immediately placed in cleaned polyethylene plastic bags, properly labeled, and transported to a dust free laboratory, where the samples were sequentially washed with acetone and ethanol, and thoroughly rinsed with deionized water to remove pharmaceutical products, particulates, and other exogenous materials. The washed hair samples were subsequently dried in an oven at 90°C for approximately 20 minutes and stored in pre-nitric acid washed Teflon sample containers. Approximately 1.0 g of dried hair sample from each participant was accurately weighed and digested with 15 mL HNO_3 (trace metal grade, purity, 99.999%) in a digestion flask. Hair sample digestions were initiated at relatively low temperatures to prevent a violent reaction until all the hair samples were fully dissolved in the nitric acid. The temperature was then subsequently increased until the solution turned pale yellow to ensure complete sample digestion. The nitric acid digested hair samples were filtered into a 25 mL volumetric flask using a Whatman filter paper (ashless) and diluted to the mark with deionized water.

2.2. Calibration Curve, Trace Element Analysis, Method Validation, and Quality Assurance

Working range standard solutions used to construct calibration curves for each element were prepared by serial dilution of 1000 ppm standard stock solutions of each element purchased from Fisher Scientific. The calibration curve for each element was constructed by plotting the absorbance obtained from FAAS analysis of the standard solution versus the element concentration. The nitric acid digested hair samples were subjected to trace element analysis using a flame atomic absorption spectrophotometer (Shimadzu, AA-6300) using a pre-mixed burner air-acetylene flame. The flow rates of the fuel and oxidant gases were always carefully optimized for each metal analysis. Other routine instrumental checks and calibrations were always performed on the spectrometer before use to ensure the accuracy, reliability and consistent performance of the spectrometer. Each sample was analyzed in triplicate and the averages of the trace element concentrations in the hair samples were calculated using the constructed calibration curves and listed in Table 1. All necessary precautions were also observed to ensure the accuracy of the results of this study. First, the acetone and ethanol used for the hair wash were of spectroscopic grade. Second, the highest purity HNO_3 acid (purity, 99.999%) was used for the hair digestion and standard solution preparations. Also, all glassware including the sample containers, digestion flasks, and volumetric flasks were pre-soaked in 6M nitric acid for three days and thoroughly rinsed with deionized water before use to remove impurities and contaminants. Additionally, all trace element sample analyses were blank subtracted.

A recovery study was also performed for each metal to further evaluate the accuracy and reliability of the results of the heavy metal concentrations obtained in the analysis. The recovery study was performed by randomly spiking ten previously analyzed hair samples with a known concentration of metal standard solution. The spiked samples were then subjected to HNO_3 digestion and metal analysis procedures under the same experimental conditions as previously used for hair sample analysis. The recovery of the metals in the spiked samples was evaluated by comparing the known concentration of the spiked metal with the concentration detected using FAAS spectrometer.

2.3. Multivariate Analysis: Regression and Principal Component Analysis

Linear regression and multivariate principal component analysis for pattern recognition of the levels of the metals in human scalp hair was performed using chemometric software (The Unscrambler, CAMO Inc., 9.4).

3. RESULTS AND DISCUSSION

3.1. Calibration Curves

Table 1 presents the results of the calibration curve parameters constructed for the investigated trace elements in the hair samples. The table indicates the wavelength used for the AAS elemental analysis, the limits of detection (LOD), and limits of quantitation (LOQ) of the trace elements analyzed. The LOD, which is defined as the minimum detectable amount of metal, was calculated using the equation: LOD = 3 s/m; where s is the signal of the blank and m is the slope of the calibration curve. The limit of quantitation, defined as the lowest measurable concentration of analytes (trace elements in this study), was evaluated using the formula: LOQ = 10 s/m. The parameters used to calculate LOQ were as previously defined for LOD. The high values of the correlation coefficients (r) obtained in Table 1 demonstrate good linear correlation of the absorbance with trace element concentrations.

3.2. Overall Trace Element Concentration in Hair Samples

The overall mean, maximum, and minimum of each trace element concentrations in hair samples are shown in Table 2. The trace of the element concentrations varied widely in the hair samples as demonstrated by the large range of trace element concentrations. The order of the mean trace element concentrations in the hair sample is Ca > Zn > Mg > Fe > Cu > Cr. Overall mean concentrations of Ca (710 μg/g), Zn (190 μg/g), Mg (120 μg/g), Fe (25 μg/g), Cu (12 μg/g), and Cr (0.20 μg/g) were found in the hair samples. Similar trace element concentrations found in this study have been reported in human scalp hair in Pakistan, Russia, Spain, Tanzania, and China [36,47,52,55, 56]. The levels of Cd and Pb concentrations were below the detection limits (<0.030 μg/g) of the instrument in the hair samples. Unquestionably, the use of a more sensitive analytical instrument such as a graphite atomic absorption or an inductively coupled plasma atomic spectrometer will be required for Cd and Pb detection at ultra-low concentrations.

3.3. Trace Element Concentration by Race and Sex

The populations were subdivided into two race categories, the Caucasians (majority) and Non-Caucasians (minority) comprising of African Americans, Hispanics, and Asians. The trace element concentrations by race and sex are shown in Tables 3 and 4, respectively. Overall mean levels of Fe (27 μg/g), Ca (750 μg/g), Mg (140 μg/g), Zn (200 μg/g), Cu (14 μg/g), and Cr (0.18 μg/g) were obtained in Caucasians. In contrast, corresponding mean levels of Fe (22 μg/g), Ca (631 μg/g), Mg (87 μg/g), Zn (163 μg/g), Cu (6.1 μg/g), Cr (0.24 μg/g) were obtained in non-Caucasians. To evaluate the differences between the mean trace element concentrations obtained in Caucasians and non-Caucasians, the mean trace element concentrations obtained for the two race categories were subjected to a statistical student t-test analysis. At a 90% confidence level, there were no significant differences ($t < 1.645$, $p = 0.01$) between the trace element concentrations obtained between Caucasians and non-Caucasians. However, at a 50% confidence level, there were significant differences between the mean concentration of Ca (t_{cal} 0.84, t_{tab} 0.674), Zn (t_{cal}, 1.602, t_{tab} 0.674), Cu (t_{cal}, 1.50, t_{tab} 0.674), and Mg ($t_{cal} = 1.574$, $t_{tab} = 0.674$) obtained in Caucasians and non-Caucasians.

Table 1. Linear regression equation, linear correlation coefficient (R) of calibration curves, wavelength of detection, limit of detection (LOD), and limit of quantitation (LOQ) obtained for each element.

Element	Regression equation	R	Wavelength (nm)	LOD (µg/g)	LOQ (µg/g)
Fe	y = 0.0040x + 0.0048	0.999	248.3	0.080	0.25
Cd	y = 0.087x + 0.29	0.973	228.8	0.030	0.11
Ca	y = 0.026x + 0.0014	0.999	422.7	0.010	0.040
Zn	y = 0.080x + 0.77	0.922	213.9	0.030	0.13
Cu	y = 0.092x + 0.055	0.999	309.93	0.030	0.12
Pb	y = 0.0097x + 0.0029	0.998	283.3	0.030	0.10
Mg	y = 0.060x + 1.6	0.998	285.2	0.050	0.16
Cr	y = 0.0065x + 0.0048	0.981	357.9	0.050	0.16

Table 2. Overall average of trace element concentration in human hair.

	Concentration (μg/g)							
	Fe	Ca	Cd	Zn	Cu	Mg	C	Pb
Average	25	710	<0.03	190	12	120	0.20	<0.030
Maximum	460	3900	<0.03	800	140	710	4.5	<0.030
Minimum	6.2	2.1	<0.03	0	0	0	0	<0.030

Trace element concentrations were also found to be gender dependent. Generally, the mean levels of all trace elements were significantly higher in females compared to males. The mean Fe level of 32 μg/g found in females was approximately twice the average level of Fe (17.80 μg/g) obtained in males. Also, the mean level of Mg (210 μg/g) obtained in females was 7 time higher than the corresponding mean Mg (27) level found in males. Also, the mean levels of Zn (210 μg/g), Cu (19 μg/g), and Cr (0.31 μg/g) were considerably higher than the mean levels of Zn (170 μg/g), Cu (4.0 μg/g), and Cr (0.080 μg/g), respectively obtained in males. Interestingly, at the 90% confidence levels, the concentration of Fe, Ca, Zn, Cu, Mg, and Cr obtained in males were significantly and statistically different ($t > 1.645$) from the corresponding level of Fe, Ca, Zn, Cu, Mg, and Cr obtained in females. At the 95% confidence level, the concentration of Ca (t_{cal}, = 6.618, t_{tab} 1.960), Zn (t_{cal}, 2.047, t_{tab} 1.960), Cu (t_{cal}, 3.107, t_{tab} 1.960), and Mg (t_{cal}, 24, t_{tab} 1.960) obtained in males were also considerably different from the corresponding level found in women, demonstrating the influence of sex on the trace element concentration in human subjects.

3.4. Effect of Smoking Habits on Trace Element Concentration

A summary of the levels of trace element concentrations found in the scalp hair of smokers and non-smokers are shown in Table 5. In general, the mean levels of Ca (550 μg/g) and Mg (77 μg/g) in smokers were less than the mean levels of Ca (740 μg/g) and Mg (130 μg/g) found in non-smokers. Also, the levels of Cu (2.5 μg/g), Zn (170 μg/g), and Cr (0.14 μg/g) found in smokers were relatively less than the corresponding levels of Cu (13 μg/g), Zn (190 μg/g), and Cr (0.21 μg/g) found in nonsmokers. Other studies have reported similar low concentrations of Ca, Mg, Cu, and Zn in smokers scalp hair compared to non-smokers in Poland [50]. However, at 95% confidence level, the mean level of Fe (40 μg/g) found in smokers in this study was statistically and significantly higher than the corresponding mean Fe level of 23 μg/g obtained in non-smokers (t_{cal} 2.12, t_{tab} 1.960).

Table 3. Trace elements concentration by race.

	Concentration (μg/g)							
	Fe	Ca	Cd	Zn	Cu	Mg	Cr	Pb
	Caucasian (n = 87)							
Average	27	750	<0.030	200	14	140	0.18	<0.030
Maximum	460	3400	<0.030	820	140	710	3.7	<0.030
Minimum	6.9	34	<0.030	0	0	0	0	<0.030
	Non-Caucasian (n = 38)							
Average	22	631	<0.030	163	6.1	87	0.24	<0.030
Maximum	170	3900	<0.030	567	61	639	4.5	<0.030
Minimum	6.2	2.1	<0.030	0	0	0	0	<0.030

Table 4. Trace element concentration by sex.

	Concentration (μg/g)							
	Fe	Ca	Cd	Zn	Cu	Mg	Cr	Pb
	Male (n = 60)							
Average	18	330	<0.030	167	4.034	27	0.080	<0.030
Maximum	170	1900	<0.030	820	51.82	230	3.2	<0.030
Minimum	6.6	2.1	<0.030	0	0	0	0	<0.030
	Female (n = 65)							
Average	32	1100	<0.030	210	19	210	0.31	<0.030
Maximum	460	3900	<0.030	560	140	710	4.5	<0.030
Minimum	6.2	36	<0.030	0	0	0	0	<0.030

3.5. Trace Element Concentration by Age

To evaluate the effect of age on trace element concentration, the sample populations were subdivided into two age group categories. The first age group category consists of people between 1 - 20 years old (n = 29), while the second age group category is made up of people 21 years and older (n = 96). Table 6 shows the results of the trace element concentrations by age. The concentrations of Fe ranged between 6.9 μg/g and 220 μg/g in the 1 - 20 years old age group while the concentrations of Fe ranged between 6.155 μg/g and 460 μg/g in people 21 years and older. There was no significant difference between the mean Fe level of 26 μg/g obtained in 1 - 20 years old age group and mean Fe level of 25 μg/g obtained in people 21 years and older. However, at the 95% confidence level, a considerably higher mean Ca concentration of 800 μg/g was obtained in 21 people years and older as compared to a relatively low average Ca level of 400 μg/g obtained in the 1 - 20 years old age group (t_{cal} 2.686, t_{tab} 1.965). In addition, the mean level of Mg (142.4 μg/g) in older people is significantly higher than the mean Mg level of 50.11 obtained in the 1 - 20 years old age group (t_{cal} = 2.780, t_{tab} = 1.965). In addition, the level of Cu (19.61 μg/g) found in 1 - 20 years old age group was considerably higher than the Cu (9.3 μg/g) found in people 21 years and older (t_{cal} = 2.27, t_{tab} = 1.965). In contrast, there was no statistical difference between mean levels of Zn (200 μg/g) mean levels of Zn (180 μg/g) found in people 21 years and older (t_{cal} = 0.863, t_{tab} = 1.965). Also, there was no significant different between the mean level of Cr of 0.21 μg/g and 0.20 μg/g, respectively, obtained in the 1 - 20 years old age group and older people (t_{cal} = 0.034, t_{tab} = 1.965). Results of the trace element concentrations in this study are similar to the trace element concentrations reported in young people in Spain [55].

Table 5. Effect of smoking habit of trace element concentration.

	Concentration (μg/g)							
	Fe	Ca	Cd	Zn	Cu	Mg	Cr	Pb
Smokers (n = 18)								
Average	40	550	<0.030	170	2.5	77	0.14	<0.030
Maximum	460	1600	<0.030	600	13	330	1.5	<0.030
Minimum	7.5	2.1	<0.030	0	0	0	0	<0.030
Non-smokers (n = 107)								
Average	23	740	<0.030	190	13	130	0.21	<0.030
Maximum	220	3900	<0.030	820	140	710	4.5	<0.030
Minimum	6.2	31	<0.030	0	0	0	0	<0.030

3.6. Ca/Mg and Zn/Cu Ratio

The ratios of Ca/Mg and Zn/Cu in hair samples are often used to evaluate Ca absorption and Cu utilization in humans. The ratios are also used as markers for various health related issues including, atherosclerosis, hypertension diabetics, insulin sensitivity, and pancreatic cancer. The utilization and absorption of Ca is highly dependent on Mg concentration (Toba et al., 1999). In general, a low Ca/Mg ratio is preferred for proper bodily function and Ca metabolism. An ideal normal Ca/Mg ratio of 2 in humans has been suggested. At elevated Mg concentrations, the absorption of Ca is severely hindered [58]. More importantly, a high Ca/Mg concentration ratio has been associated with various health issues, including, atherosclerosis and hypertension [59,60]. Additionally, Ca/Mg imbalances may lead to other health related issues such as diabetics and alterations in insulin sensitivity [61] necessitating the need to critically evaluate the Ca/Mg ratio in humans. Table 7 shows the Ca/Mg ratio in the hair samples to be 5.9. The calculated Ca/Mg ratio was also found to depend on age, sex, race, and smoking habits. Generally, the Ca/Mg ratio in older people is less than the Ca/ Mg ratio found in younger people. Also, non-Caucasians have a relatively larger Ca/Mg ratio of 7.2 compared to a Ca/Mg ratio of 5.5 found in Caucasians. The Ca/Mg ratio (12) found in males was twice the corresponding Ca/Mg ratio of 5.1 found in females. Smoking also seems to influence the Ca/Mg ratio. The Ca/Mg ratio (7.2) in smokers is larger than the Ca/Mg ratio of 5.8 found in nonsmokers. Other studies have also reported similarly higher Ca/Mg ratios in hair samples of smokers compared to non-smokers [50].

Accurate knowledge of Zn to Cu concentration ratios in humans is also of critical health importance because a high Zn/Cu ratio may influence the utilization of Cu, resulting in Cu deficiency in the body. For example, reduced erythrocyte copper-zinc superoxide dismutase activity at elevated Zn plasma concentrations has been reported [62]. Additionally, elevated Zn concentrations have been linked to significant reductions in Cu plasma caeruloplasmin, serious anemia, and alterations in immune responses and serum lipids [63-65]. Besides, an increase in the Zn/Cu ratio has been observed in pancreatic cancer patients compared to healthy humans [66].

Table 6. Trace elements concentration by age group.

	Concentration (μg/g)							
	Fe	Ca	Cd	Zn	Cu	Mg	Cr	Pb
Age 1 - 20 (n = 29)								
Average	26	400	<0.030	200	20	50	0.21	<0.030
Maximum	220	1300	<0.030	360	140	300	3.2	<0.030
Minimum	6.9	36	<0.030	37	0	0	0	<0.030
Age 21 and older (n = 96)								
Average	25	800	<0.030	180	9.3	140	0.20	<0.030
Maximum	460	3900	<0.030	820	130	710	4.5	<0.030
Minimum	6.2	2.1	<0.030	0	0	0	0	<0.030

The Zn/Cu concentration ratios in hair samples by age group, sex, race, and smoking habit are presented in Table 7. A higher Zn/Cu concentration ratio of 69 was found in smokers compared to Zn/Cu ratio of 14 found in non-smokers. The Zn/Cu ratio of 20 found in older people was also significantly larger than the Zn/Cu ratio of 10 observed in younger people. However, Caucasians have relatively low Zn/Cu ratio (14) compared to the Zn/ Cu ratio of 27 found in non-Caucasians. Also, a female's scalp hair contained a low Zn/Cu ratio of 11 compared with a high Zn/Cu ratio of 40 found in male scalp hair samples.

3.7. Analytical Method Validation and Recovery Study

Commercial human hair trace element reference standards were not readily available, consequently a spike recovery was performed to evaluate the accuracy of the results obtained from this study. The results of the recovery study performed for each metal analysis in the hair samples are presented in Table 8. Overall, the percentage recovery of the metal analysis in the hair samples ranged between 95% and 104%. The percent recovery obtained in this study falls within the normal acceptable range of 90% - 110% for a good recovery study. The high percentage recovery obtained from the study validates the accuracy of the method and the reliability of the levels of metal concentration in this study. In addition, the obtained high recovery demonstrates insignificant loss or addition of metals to the hair samples during the sample preparation, sample digestion or instrumental metal analysis procedures.

Table 7. Ca/Mg and Zn/Cu.

Category	Ca/Mg ratio	Zn/Cu ratio
Younger people (1 - 20 years age group)	8.1	10
Older people (older than 21 years age group)	5.6	20
Caucasian	5.5	14
Non-Caucasian	7.2	27
Smokers	7.2	70
Non-smokers	5.8	14
Males	12	40
Females	5.1	11
Overall	5.9	16

3.8. Linear Regression and Principal Component Analysis

To further gain a better insight and understanding of the inter-element association of trace elements in the scalp hair samples, the levels of trace elements were subjected to a linear regression and principal component analysis (PCA). Table 9 presents the summary of the correlation coefficients (r) of the inter-element associations in the hair samples. Generally, the levels of the trace elements in hair samples were poorly correlated. However, significant correlations were found between Ca and Mg ($r = 0.840$, $p = 0.05$). Poor correlation between the levels of trace elements in hair suggests that there may be no particular common source of the trace elements in the hair.

Diet, eating habits, and the use of food supplements varies widely among people; therefore, people may obtain the required trace elements from diverse sources, resulting in the poor correlation of trace elements in hair samples. Similar poor inter-element associations in hair samples have also been observed and reported elsewhere [66].

Table 8. Recovery study for FAAS metal analysis in hair samples.

Metal	% Recovery
Pb	96.0
Cd	94.6
Fe	97.8
Ca	102.4
Mg	98.1
Zn	95.2
Cu	104.1
Cr	102.4

Table 9. Correlation coefficient (r) values of the association between trace elements in human hair samples.

	Fe	Ca	Mg	Zn	Cu	Cr
Fe		0.171	0.092	0.035	0.037	0.154
Ca	0.171		**0.840**	0.174	0.105	0.044
Mg	0.092	**0.840**		0.217	0.102	0.055
Zn	0.035	0.174	0.217		0.082	0.050
Cu	0.037	0.105	0.102	0.082		−0.042
Cr	0.154	0.044	0.055	0.050	−0.042	

Principal component analysis is a modern chemometric approach often employed for pattern recognition in a complex data set. PCA is often used to obtain hidden information that is not apparent from a conventional data analysis. Detailed mathematical discussions and procedures of PCA have been widely reported [67-69] and is beyond the scope of this study. In brief, PCA is usually performed by transforming the original data set or matrix from the initial coordinate system made up of n-variables into a new orthogonal coordinate system. The new variable space is typically composed of a smaller number of orthogonal vectors known as principal components (PCs). For example, a set of data with n-variables ($n = 6$ in this study, corresponding to the number of trace elements determined in hair samples) will require an n-dimensional coordinate system, known as variable space to represent it. The most vital assumption of multivariate analysis is that the directions with maximum variance in a data set are associated with useful information. Consequently, the first principal component (PC1) is drawn through the variable space in the direction of the maximum variance in of the original data set. The second principal component (PC2) is then drawn through the data set in the direction of the next largest variance, so that PC1 and PC2 are orthogonal to each other. The third principal component (PC3) is subsequently drawn so that PC1, PC2, and PC3 are all orthogonal to each other. Thus, PC1, PC2, and PC3 now represent a set of variance-scaled eigenvectors that provide a new orthogonal coordinate system on which the original data set can be represented. The individual principal components can be calculated as shown in Equation (1). Each PC is a linear combination of the n-vectors that make up the original variable space.

$$PC1 = P_{11}X_1 + P_{21}X_2 + P_{31}X_3 + L + P_{n1}X_n \quad (1)$$

where, PC1 is the first principal component and X_1, X_2, X_3, … X_n, are the magnitudes of the levels of trace elements in this study, and the coefficients (P_{Ij}) of the linear combination are called x-loadings. The coordinates of the data on the new PCs coordinate system (t_i) are referred to as scores. The magnitude of the scores reveals the importance of PC_i to that data point. Groupings of data points on the scores plot may reveal obscured structure in the data. Consequently, new insight is often gained as new relationships that were formerly obscured in the old coordinate system are revealed, allowing pattern recognition in a data set. The use of PCA for data analysis is desirable because it often reduces the dimensionality of the data set to a much smaller value.

The results of the PCA shows that the first two PCs accounted for 98% of the variability in the levels of trace element in the hair samples. Therefore, two PCs (PC1 and PC2) are adequate to represent the data set. Equations (2) and (3) show a mathematical expression for PC1, and PC2, respectively obtained from the hair analysis in this study.

$$PC2 = -5.165e^{-3}[Fe] - 6.477e^{-2}[Ca] + 0.979[Zn] + 1.659e^{-2}[Cu] + 0.195\ [Mg] + 2.770e^{-4}[Cr] \quad (2)$$

$$PC2 = -5.165e^{-3}[Fe] - 6.477e^{-2}[Ca] + 0.979[Zn] + 1.659e^{-2}[Cu] + 0.195\ [Mg] + 2.770e^{-4}[Cr] \quad (3)$$

Figure 1(a) is the x-loading of the plot of PC1 versus the PC2. While all the trace elements are important, calcium and magnesium were found to be the most significant elements, contributing to the PC1, with x-loading values of 0.98 and 0.18, respectively. However, Zn and Mg were found to be the major contributors to PC2, with x-loading values of 0.98 and 0.20 respectively. It is very interesting to note that, the overall calculated Ca/Mg ratio value of 5.87 in Table 7 was similar to the corresponding x-loading Ca/Mg ratio of 5.3 obtained from PCA. However, the x-loading Zn/Cu ratio of 59 obtained from PCA was significantly larger than the corresponding overall Zn/Cu ratio value of 16.0 obtained in Table 7. These differences may be associated with a wider range of Zn/Cu ratio (69 - 10) obtained for different categories of people in Table 7. Nontheless, further critical observation of the data revealed that the x-loading Zn/Cu ratio of 59 was similar to the average Zn/Cu ratio value of 54 obtained in smokers (69) and in men (40). With further study and more data analysis, the ratio of the x-loadings value of Ca/Mg and Zn/Cu from PCA can therefore be potentially used for rapid screening of calcium and copper utilization and for medical diagnosis of people for suseptiability to atherosclerosis, hypertension diabetics, insulin sensitivity, and pancreatic cancer.

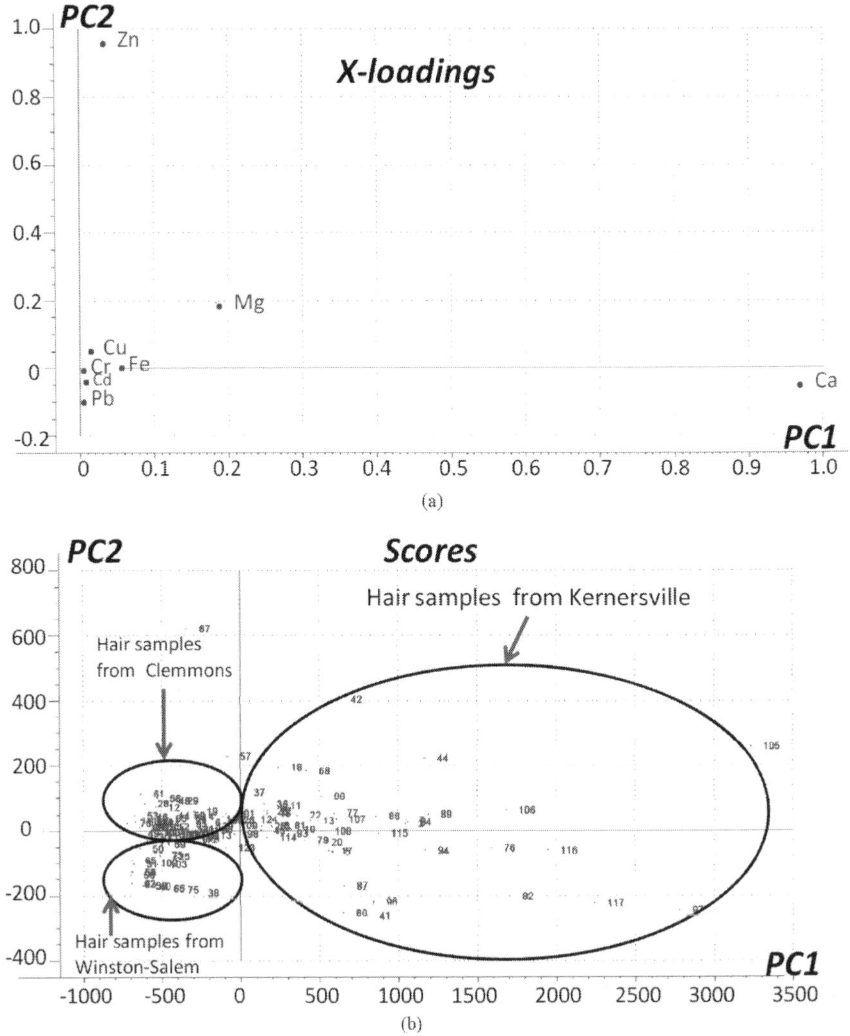

Figure 1. (a) X-loadings of the plot of PC2 versus PC1; (b) Scores plot the second principal component (PC2) versus the first principal component (PC1).

Figure 1(b) is the scores plot of the PC1 versus PC2. A closer examination of the plot reveals an interesting pattern with groupings of hair samples based on the locations where the hair samples were collected. The numbers in the plots are the different number codes assigned to different hair samples. For example, hair samples collected in

Kernersville were grouped together in the first and second quadrants in Figure 1(b). However, hair samples collected in Winston-Salem and Clemmons were similarly grouped together in the lower left of the figure in the third and fourth quadrant, respectively, in Figure 1(b). Other studies have reported the practical application of PCA in elemental analysis for pattern recognition to differentiate and ascertain the origin of several agricultural products [70,71].

4. CONCLUSION

The results of the flame atomic absorption spectroscopic determination of trace elements in human scalp hair were reported. The levels of trace elements in hair samples were further subjected to linear regression and principal component analysis for data analysis and pattern recognition. The result of this study showed wide variation in trace element concentrations in human scalp hair which may be attributed to differences in eating habits. In addition, trace element concentrations were found to be highly dependent on age, sex, race, and smoking habits. Cadmium and Pb were below the detection limit of 0.030 μg/g in the hair samples. Overall, the levels of trace element concentrations in this study were within the expected normal range of healthy humans and similar to reported levels of trace element concentrations in healthy individuals elsewhere. Future study will include comparative analysis and determination of trace element concentrations in human scalp hair of cancer, diabetics, and obesity patients. The results of these findings will be communicated in future manuscripts.

ACKNOWLEDGEMENTS

The authors acknowledge financial support from National Science Foundation HBCU-UP Award # HRD- 0927905, US Department of Education, Title III grant to Winston Salem State University.

REFERENCES

1. K. M. Hambidge, "Trace Elements in Human and Animal Nutrition," 5th Edition, Academic Press, San Diego, 1987.

2. World Health Organization, "Trace Elements in Human Nutrition and Health," WHO, Geneva, 1996.

3. Institute of Medicine, "Food and Nutrition Board. Dietary Reference Intakes for Vitamin A, Vitamin K, Arsenic, Boron, Chromium, Copper, Iodine, Iron, Manganese, Molybdenum, Nickel, Silicon, Vanadium and Zinc," National Academy Press, Washington DC, 2001.

4. M. Luno, P. Roncales and D. Djenane, "Beltran, Beef Shelf Life in Low O_2 and High CO_2 Atmospheres Containing Different Low CO Concentrations," Meat Science, Vol. 55, No. 4, 2000, pp. 413-419.

5. M. Seyfert, R. A. Mancini, M. C. Hunt, J. Tang, C. Faustman, "Influence of Carbon monoxide in Package Atmospheres Containing Oxygen on Color, Reducing Activity, and Oxygen Consumption of Five Bovine Muscles," Meat Science, Vol. 75, No. 3, 2007, pp. 432-442.

6. D. Voet, J. G. Voet and C. W. Pratt, "Fundamentals of Biochemistry: Life at the Molecular Level," 3rd Edition, John Wiley & Sons, Inc., Hoboken, 2008.

7. J. Y. Jeong and J. R. Claus, "Color Stability and Reversion in Carbon Monoxide Packaged Ground Beef," Meat Science, Vol. 85, No. 3, 2010, pp. 525-530.

8. H. Tapiero, L. Gate and K. D. Tew, "Iron: Deficiencies and Requirements," Biomedicine & Pharmacotherapy, Vol. 55, No. 6, 2001, pp. 324-332.

9. S. Huang, Y. Yang, C. Cheng, J. Chen and C. Lin, "The Etiology and Treatment Outcome of Iron Deficiency and Iron Deficiency Anemia in Children," Journal of Pediatric Hematology/Oncology, Vol. 32, No. 4, 2010, pp. 282-285.

10. N. Fabris and F. Mocchegiani, “Zinc, Human Diseases and Aging,” Aging Clinical and Experimental Research, Vol. 7, No. 2, 1995, pp. 77-93.

11. G. Vivoli, M. Bergomi, P. Borella, G. Fantuzzi and E. Caselgrandi, “Cadmium in Blood, Urine and Hair Related to Human Hypertension,” Journal of Trace Elements and Electrolytes in Health and Disease, Vol. 3, No. 3, 1989, pp. 139-145.

12. J. Schwartz, “Low-Level Lead Exposure and Children’s IQ: A Meta-Analysis and Search for a Threshold,” Environmental Research, Vol. 65, No. 1, 1994, pp. 42-55.

13. S. Tong, Y. E. von Schirnding and T. Prapamontol, “Environmental, Lead Exposure: A Public Health Problem of Global Dimensions,” Bulletin of World Health Organization, Vol. 78, No. 9, 2000, pp. 1068-1077.

14. Agency for Toxic Substances and Diseases Registry, “Toxicological Profile for Lead,” US Department of Health and Human Services, Public Health Services, 2005. http://www.atsdr.cdc.gov/toxprofiles/tp13.pdf

15. G. Hotter, L. M. Fels, D. Closa, J. Rosello, H. Stolte and E. Gelpi, “Altered Levels of Urinary Prostanoids in LeadExposed Workers,” Toxicology Letter, Vol. 77, No. 1-3, 1995, pp. 309-312.

16. M. P. Nava-Hernandez, L. A. Hauad-Marroquin, S. BassolMayagoitia, G. GarciaArenas, R. Mercado-Hernandez, M. A. Echavarri-Guzman and R. M. Cerda-Flores, “Lead, Cadmium, and Arsenic-Induced DNA Damage in Rat Germinal Cells,” DNA Cell Biology, Vol. 28, No. 5, 2009, pp. 241-248.

17. Y. von Schirnding, A. Mathee, M. Kibel, P. Robertson, N. Strauss and R. A. Blignaut, “Study of Pediatric Blood Levels in a Mining Area in South Africa,” Environmental Research, Vol. 93, No. 3, 2003, pp. 259-263.

18. G. C. Lough, J. J. Schauer, J. Park, M. M. Shafer, J. T. DeMinter and J. P. Weinstein, “Emissions of Metals Associated with Motor

Vehicle Roadways," Environmental Chemistry and Technology, Vol. 39, No. 3, 2005, pp. 826- 836.

19. J. Nriagu, N. Oleru, C. Cudioe and A. Chine, "Lead Poisoning of Children in Africa, III. Kaduna, Nigeria," Science of the Total Environment, Vol. 197, No. 1-3, 1997, pp. 13-19.

20. P. C. Onianwa and S. O. Fakayode, "Lead Contamination of Topsoil and Vegetation in the Vicinity of a Battery Factory in Nigeria," Environmental Geochemistry and Health, Vol. 22, No. 3, 2000, pp. 211-218.

21. O. E. Orisakwe, "Environmental Pollution and Blood Lead Levels in Nigeria: Who Is Unexposed?" International Journal of Occupational and Environmental Health, Vol. 15, No. 3, 2009, pp. 315-317.

22. J. E. Fergusson, K. A. Hibbard and R. L. H. Ting, "Lead in Human Hair: General Survey-Battery Factory Employees and Their Families," Environmental Pollution (Ser B), Vol. 2, No. 3, 1981, pp. 235-248.

23. S. O. Fakayode and B. I. Olu-Owolabi, "Heavy Metal Contamination of Roadside Topsoil in Osogbo, Nigeria: Its Relationship to Traffic Density and Proximity to Highways," Environmental Geology, Vol. 44, No. 2, 2003, pp. 150-157.

24. S. Bakirdere and M. Yaman, "Determination of Lead, Cadmium and Copper in Roadside Soil and Plants in Elazig, Turkey," Environmental Monitoring and Assessment, Vol. 136, No. 1-3, 2008, pp. 401-410.

25. K. Suzuki, T. Yabuki and Y. Ono, "Roadside Rhododendron Pulchrum Leaves as Bioindicators of Heavy Metal Pollution in Traffic Areas of Okayama, Japan," Environmental Monitoring and Assessment, Vol. 149, No. 1-4, 2009, pp. 133-141.

26. H. W. Mielke, D. Dugas, P. W. Mielke, K. S. Smith, S. L. Smith and C. R. Gonzales, "Association between Soil and Soil Lead and Childhood Blood Lead in Urban New Orleans and Rural Lafourche

Parish of Louisiana," Environmental Health Perspectives, Vol. 105, No. 4, 1997, pp. 950-954.

27. U. Ranft, T. Delschen, M. Machtolf, D. Sugiri and M. Wilhelm, "Lead Concentration in the Blood of Children and Its Association with Lead in Soil and Ambient Air. Trends between 1983 and 2000 in Duisburg," Journal of Toxicology and Environmental Health, Part A, Vol. 71, No. 11-12, 2008, pp. 710-715.

28. A. Kumar and P. Pastore, "Lead and Cadmium in Soft Plastic Toys," Current Science India, Vol. 93, No. 6, 2007, pp. 818-822.

29. European Standard EN 71-3, "Safety of Toys-Part 3, Migration of Certain Elements," 1994.

30. Consumer Product Safety Commission, "Determination of Safe Levels of Lead in Paints," Federal Register, Vol. 42, No. 32, 1977, pp. 9404-9406.

31. A. Leibovitz, E. Lubart, J. Wainstein, Y. Dror and R. Segal, "Serum Trace Elements in Elderly Frail Patients with Oropharyngeal Dysphagia," Journal of Nutritional Science and Vitaminology, Vol. 55, No. 5, 2009, pp. 407-411.

32. B. Gulson, K. Mizon, A. Taylor, M. Korsch, J. Stauber, J. M. Davis, H. Louie and M. A. L. Wu, "Longitudinal Monitoring of Selected Elements in Blood of Healthy Young Children," Journal of Trace Elements in Medicine and Biology, Vol. 22, No. 3, 2008, pp. 206-214.

33. J. L. Rodrigues, B. L. Batista, M. Fillion, C. J. S. Passos, D. Mergler and F. Barbosa, "Trace Element Levels in Whole Blood of Riparian Villagers of The Brazilian Amazon," Science of the Total Environment, Vol. 407, No. 13, 2009, pp. 4168-4173.

34. M. L. Carvalho and A. F. Marques, "X-Ray Fluorescence Spectrometry: Applications in Trace Elements Studies in Human Tissues from Patients with Cirrhosis," X-Ray Spectrometry, Vol. 30, No. 6, 2001, pp. 397-402.

35. A. O. Ogunfowokan, J. P. Kaisam and M. O. Balogun, "Study of Trace Elements in Urine of Some Nigerian Medical Patients," Toxicological and Environmental Chemistry, Vol. 91, No. 3, 2009, pp. 435-449.

36. Q. Pasha, S. A. Malik, J. Iqbal and M. H. Shah, "Characterization and Distribution of the Selected Metals in the Scalp Hair of Cancer Patients in Comparison with Normal Donors," Biological Trace Elemment Research, Vol. 118, No. 3, 2007, pp. 207-216.

37. R. Pereria, R. Riberio and F. Goncalves, "Scalp Hair Analysis as a Tool in Assessing Human Exposure to Heavy Metals (S. Domingos Mine, Portugal)," Science of the Total Environment, Vol. 327, No. 1-3, 2004, pp. 81-92.

38. M. Nadal, A. Bocio, M. Schuhmacher and J. L. Domingo, "Monitoring Metals in Population Living in the Vicinity of Hazardous Waste Incinerator: Levels in Hair of School Children," Biological Trace Elemment Research, Vol. 104, No. 3, 2005, pp. 203-214.

39. S. Majumdar, J. Chatterjee and K. Chaudhuri, "Ultrastructural and Trace Metal Studies on Radiographers' Hair and Nails," Biological Trace Element Research, Vol. 67, No. 2, 1997, pp. 127-138.

40. M. J. Welch, L. T. Sniegoski, C. C. Allgood and M. Habram, "Hair Analysis for Drugs of Abuse: Evaluation of Analytical Methods, Environmental Issues, and Development of Reference Materials," Journal of Analytical Toxicology, Vol. 17, No. 7, 1993, pp. 389-398.

41. M. Balikova, "Hair Analysis for Drugs of Abuse: Plausibility of Interpretation," Biomed Pap Med Fac Univ Palacky Olomouc Czech Republic, Vol. 149, No. 2, 2005, pp. 199-207.

42. H. I. Afridi, T. G. Kazi, N. Kazi, G. A. Kandhro, J. A. Baig, M. K. Jamali, M. B. Arain and A. Q. Shah, "Interactions between Cadmium and Zinc in the Biological Samples of Pakistani Smokers

and Nonsmokers Cardiovascular Disease Patients," Biological Trace Elemment Research, Vol. 139, No. 3, 2011, pp. 257-268.

43. F. L. Barbieri, A. Cournil, J. E. S. Sarkis, E. Bénéfice and J. Gardon, "Hair Trace Elements Concentration to Describe Polymetallic Mining Waste Exposure in Bolivian Altiplano," Biological Trace Elemment Research, Vol. 139, No. 1, 2011, pp. 10-23.

44. E. Rębacz, I. Baranowska-Bosiacka and D. Chlubek, "The Content of Selected Chemical Elements in the Hair of Young Men of the Bantu Language Group from Tanzania versus Environmental and Social Conditioning," Biological Trace Element Research, Vol. 137, No. 3, 2010, pp. 262-279.

45. V. L. Dressler, D. Pozebonb, M. F. Meskoc, A. Matuschd, U. Kumtabtime, B. Wue and J. S. Beckere, "Biomonitoring of Essential and Toxic Metals in Single Hair Using On-line Solution-Based Calibration in Laser Ablation Inductively Coupled Plasma Mass Spectrometry," Talanta, Vol. 82, No. 5, 2010, pp. 1770-1777.

46. I. A. Olusegun and N. S. Gerhard, "Lead Exposure: A Contributing Cause of the Current Breast Cancer Epidemic in Nigerian Women," Biological Trace Elemment Research, Vol. 136, No. 2, 2010, pp. 127-139.

47. S. Zaichick and V. Zaichick, "The Effect of Age and Gender on 37 Chemical Element Contents in scalp Hair of Healthy Humans," Biological Trace Elemment Research, Vol. 134, No. 1, 2010, pp. 41-54.

48. G. R. Lloyd, S. Ahmad, M. Wasimc and R. G. Brereton, "Pattern Recognition of Inductively Coupled Plasma Atomic Emission Spectroscopy of Human Scalp Hair for Discriminating between Healthy and Hepatitis C Patients," Analytica Chimica Acta, Vol. 649, No. 1, 2009, pp. 33-42.

49. T. Wang, J. Fu, Y. Wang, C. Liao, Y. Tao and G. Jiang, "Use of Scalp Hair as Indicator of Human Exposure to Heavy Metals in an

Electronic Waste Recycling Area," Environmental Pollution, Vol. 157, No. 8-9, 2009, pp. 2445-2451.

50. A. Unkiewicz-Winiarczyk, A. Bagniuk, K. GromyszKałkowska and E. Szubartowska, "Calcium, Magnesium, Iron, Zinc and Copper Concentration in the Hair of Tobacco Smokers," Biological Trace Elemment Research, Vol. 128, No. 2, 2009, pp. 152-160.

51. T. Chao, C. Hui and X. Chengyun, "The Prediction of Cardiovascular Disease Based on Trace Element Contents in Hair and A Classifier of Boosting Decision Stumps," Biological Trace Elemment Research, Vol. 129, No. 1-3, 2009, pp. 9-19.

52. N. K. Mohammed and N. M. Spyrou, "Determination of Trace Elements in Hair from Tanzanian Children: Effect of Dietary Factors," Journal of Radioanalytical and Nuclear Chemistry, Vol. 278, No. 2, 2008, pp. 455-458.

53. A. Rahman, M. A. K. Azad, I. Hossain, M. M. A. S. Qusar, W. Bari, F. Begum, S. M. I. Huq and A. Hasnat, "Zinc, Manganese, Calcium, Copper, and Cadmium Level in Scalp Hair Samples of Schizophrenic Patients," Biological Trace Elemment Research, Vol. 127, No. 2, 2009, pp. 102-108.

54. J. L. Rodrigues, B. L. Batista, J. A. Nunes, C. J. S. Passos and F. Barbosa Jr., "Evaluation of the Use of Human Hair for Biomonitoring the Deficiency of Essential and Exposure to Toxic Elements," Science of the Total Environment, Vol. 405, No. 1-3, 2008, pp. 370-376.

55. M. J. González-Muñoz, A. Peña and I. Meseguer, "Monitoring Heavy Metal Contents in Food and Hair in a Sample of Young Spanish Subject," Food Chemistry and Toxicology, Vol. 46, No. 9, 2008, pp. 3048-3052.

56. Y. Li, L. Yang, W. Wang, H. Li, J. Lv and X. Zou, "Trace Element Concentrations in Hair of Healthy Chinese Centenarians," Science of the Total Environment, Vol. 409, No. 8, 2011, pp.1385-1390.

57. U.S. Census Bureau, 2010. http://www.census.gov/

58. Y. Toba, R. Masuyam, K. Katot, Y. Takadal, S. Aoet, K. Suzuki, Y. Toba, R. Masuyama, K. Katot and Y. Takadal, "Effects of Dietary Magnesium Level on Calcium Absorption in Growing Male Rats," Nutrition Research, Vol. 19, No. 5, 1999, pp. 783-793.

59. K. Kisters, F. Wessels, H. Kuper, F. Tokmak, E. Krefting, B. Gremmler, M. Kosch, M. Barenbrock and M. Hausberg, "Increased Calcium and Decreased Magnesium Concentrations and an Increased Calcium/Magnesium Ratio in Spontaneously Hypertensive Rats versus WistarKyoto Rats: Relation to Arteriosclerosis," American Journal of Hypertension, Vol. 17, No. 1, 2004, pp. 59-62.

60. M. Kosch, M. Hausberg, G. Westermann, J. Koneke, F. Matzkies, K. H. Rahn and K. Kisters, "Alterations in Calcium and Magnesium Content of Red Cell Membranes in Patients with Primary Hypertension," American Journal of Hypertension, Vol. 14, No. 3, 2001, pp. 254-258.

61. A. Haenni, L. B. R. RetIeland, P. Anderssson, L. Lind and H. Lithell, "The Alterations In Insulin Sensitivity Angiotensin Converting Enzyme in Ii During Inhibitor Treatment are Related to Changes in The Calcium/Magnesium Balance," American Journal of Hypertension, Vol. 10, No. 2, 1997, pp. 145-151.

62. P. W. F. Fischer, A. Giroux and M. R. L'Abbk, "Effect of Zinc Supplementation on Copper Statusin Adult Man," The American Journal of Clinical Nutrition, Vol. 40, No. 4, 1984, pp. 743-746.

63. W. P. Patterson, M. Winkelmann and M. C. Perry, "Zincinduced Copper Deficiency: Megamineral Sideroblastic Anemia," Annals of Internal Medicine, Vol. 103, No. 3, 1985, pp. 385-386.

64. K. G. Porter, D. McMaster, M. E. Elmes and A. H. Love, "Anaemia and Low Serum-Copper during Zinc Therapy," Lancet, Vol. 2, No. 8041, 1977, p. 774.

65. R. K. Chandra, "Excessive Intake of Zinc Impairs Immune Responses," Journal of American Medical Association, Vol. 252, No. 11, 1984, pp. 1443-1446.

66. C. Fabris, R. Farini, G. D. Favero, G. Gurrieri, A. Piccoli, G. C. Sturniolo, A. Panucci and R. Naccarato, "Copper, Zinc and Copper/Zinc Ratio in Chronic Pancreatitis and Pancreatic Cancer," Clinical Biochemistry, Vol. 18, No. 16, 1985, pp. 373-375.

67. K. R. Beebe, R. J. Pell and M. B. Seasholtz, "Chemometrics—A Practical Guide," Wiley-Interscience, New York, 1998.

68. K. W. Busch, I. M. Swamidoss, S. O. Fakayode and M. A. Busch, "Determination of the Enantiomeric Composition of Guest Molecules by Chemometric Analysis of the UVVisible Spectra of Cyclodextrin Guest-Host Complexes," Journal of the American Chemical Society, Vol. 125, No. 7, 2003, pp. 1690-1691.

69. M. Otto, "Chemometrics," Wiley-VCH: Weinheim, Weinheim, 1999.

70. E. Marengo and M. Aceto, "Statistical Investigation of the Differences in the Distribution of Metals in Nebbiolobased Wines," Food Chemistry, Vol. 81, No. 4, 2003, pp. 621-630.

71. C. Diaz, J. E. Conde, D. Estevez, S. J. Perez Olivero and J. P. Perez Trujillo, "Application of Multivariate Analysis and Artificial Neural Networks for the Differentiation of Red Wines from the Canary Islands According to the Island of Origin," Journal of Agricultural and Food Chemistry, Vol. 51, No. 15, 2003, pp. 4303-4307.

CHAPTER 5

Structural characterization of vanadium oxide catalysts supported on nanostructured silica SBA-15 using X-ray absorption spectroscopy

Anke Walter1, Rita Herbert2, Christian Hess2,3, Thorsten Ressler1*

[1]Institut für Chemie, Technische Universität Berlin, Strasse des 17. Juni 135, 10623 Berlin, Germany.

[2]Fritz-Haber-Institute of the Max-Planck-Society, Department of Inorganic Chemistry, Faradayweg 4-6, 14195 Berlin, Germany.

[3]Eduard-Zintl-Institut für Anorganische und Physikalische Chemie, Technische Universität Darmstadt, Petersenstr. 20, 64287 Darmstadt, Germany.

ABSTRACT

The local structure of vanadium oxide supported on nanostructured SiO_2 (V_xO_y/SBA-15) was investigated by in situ X-ray absorption spectroscopy (XAS). Because the number of potential parameters in XAS data analysis often exceeds the number of "independent" parameters, evaluating the reliability and significance of a particular fitting procedure is mandatory. The number of independent parameters

(Nyquist) may not be sufficient. Hence, in addition to the number of independent parameters, a novel approach to evaluate the significance of structural fitting parameters in XAS data analysis is introduced. Three samples with different V loadings (i.e. 2.7 wt %, 5.4 wt %, and 10.8 wt %) were employed. Thermal treatment in air at 623 K resulted in characteristic structural changes of the V oxide species. Independent of the V loading, the local structure around V centers in dehydrated V_xO_y/SBA-15 corresponded to an ordered arrangement of adjacent V_2O_7 units. Moreover, the V_2O_7 units were found to persist under selective oxidation reaction conditions.

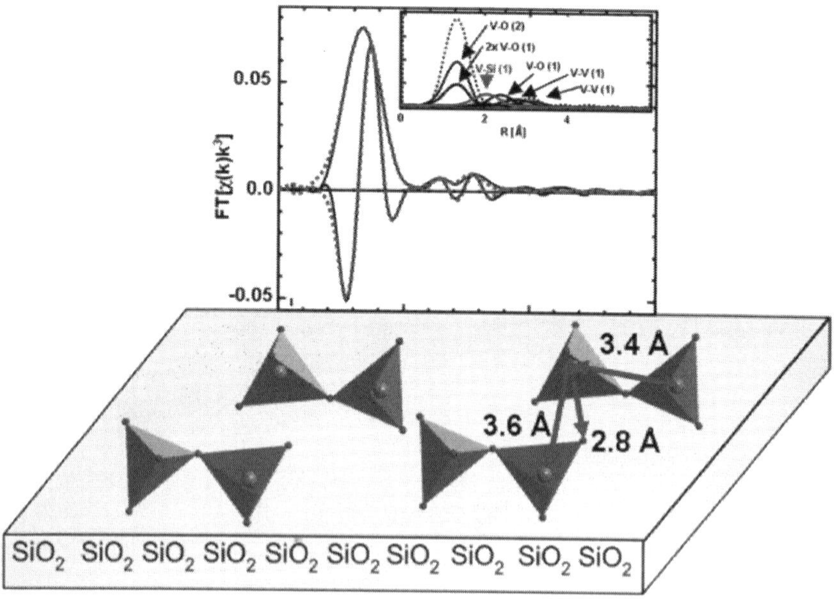

1. BACKGROUND

Mixed transitions metal oxides (e.g. MoVNbTe oxides) are active in selective oxidation of propane to acrylic acid. In contrast to various binary oxides (e.g. MoO_3 or V_2O_5), these mixed oxides exhibit a much higher selectivity. However, the origin of the promoting effect of, for instance, vanadium in mixed oxides is largely unknown. Hence, model systems are sought which enable conclusions on structure activity relationships of individual metal centers in active catalysts. For that, supported metal oxides possess two major advantages over bulk oxides. First, particular metal oxide structures which are not readily available for investigations under reaction conditions can be stabilized and studied on suitable support materials [1]. Second, dispersed supported metal oxides simplify correlating the local structure around the metal centers with their catalytic performance. Distinguishing active metal centers at the surface from metal centers in the bulk of conventional oxide catalysts is no longer required.

V_xO_y supported on SBA-15 (nanostructured SiO_2) [2] constitutes a suitable model system to investigate the role of vanadium during selective oxidation catalysis [3–6]. Structural characterization of V_xO_y supported on SiO_2 has been subject of many spectroscopic studies including IR [[7–9], XPS [6, 10, 11], Raman [10–15], UV-VIS [11, 13–16] and EXAFS [13, 17–22]. A recent review of spectroscopic investigations and structural characteristics of various supported vanadium oxides has been presented by Weckhuysen and Keller [23]. It is assumed, that the structure of supported vanadium oxide depends on both amount of vanadium and degree of hydration [14]. Hence, most studies were performed on V_xO_y/SiO_2 samples exhibiting low vanadium loading (< 10 wt %). At these loadings a monolayer of supported V_xO_y species is assumed and crystalline V_2O_5 is not detectable. Under ambient conditions the structure of hydrated vanadium oxide supported on SiO_2 resembles that of V_2O_5 [10, 13, 18]. Thermal treatment in oxygen results in dehydration of the vanadium oxide species. This dehydrated state has been proposed to consist of isolated VO_4 tetrahedrons bond to the SiO_2 support [13, 17, 24, 25]. However, V_2O_7 dimers or further extended structures supported on SiO_2 have not been excluded [15]. In total, the structure of dehydrated vanadium oxide species supported on SiO_2 remains under debate.

XAS is particularly suitable to study supported catalysts under reaction conditions. The average valence, for instance, can be readily obtained by comparison with known reference compounds. Elucidating the geometric structure, however, is often more difficult. In the conventional approach theoretical XAFS scattering amplitudes and phases are calculated for a suitable model structure. Subsequently, a sum of theoretical XAFS functions is refined to the experimental data. Structural parameters like coordination numbers, nearest neighbor distances, and disorder parameters may be determined. However, more often than not, the number of potential parameters exceeds the number of "independent" parameters. The upper limit may be calculated from Fourier theory and must not be exceeded. Nonetheless, it appears that even refinements employing a much smaller number of freely varied parameters may yield ambiguous structural results. The often used Nyquist criteria may not be sufficient to deem a fitting procedure reliable. Basically, one pair of strongly correlated parameters suffices to render a seemingly good agreement between experimental data and theoretical model structure meaningless. While this case may be clearly indicated by the correlation matrix of the refinement, other pitfalls may be less obvious. Hence, procedure are sought that enable evaluating the significance of each fitting parameter individually.

Here, we have performed in situ XAS investigations of V_xO_y supported on SBA-15 in the hydrated and dehydrated state. The same materials were already carefully characterized by several standard techniques (i.e. physisorption, TEM, IR, Raman, UV-Vis, and XPS) and the results of these studies have been described in Ref [9, 10, 14]. In particular, using UV-Vis and Raman spectroscopy, Hess et al. showed that the catalysts are similar to other systems previously described in the literature. However, the results obtained could not unequivocally determine the local structure around the V species on SiO_2, as it was also the case in previous studies. Our approach focused on elucidating the local structure around the vanadium centers in the dehydrated state of V_xO_y-SBA-15 model catalysts with different vanadium loadings. A detailed XAFS data analysis, in particular of higher V-V distances, was performed together with a detailed evaluation of the significance of the fitting parameters employed. This procedure permitted detailed conclusions on the extended local structure of the vanadium oxide species supported on SBA-15.

2. RESULTS AND DISCUSSION

2.1 Local structure of dehydrated V_xO_y/SBA-15 - Comparison to V oxide references

Characterization of pore structure and surface area, and optical spectroscopic investigations of the same model catalysts studied here have been previously described [9, 10, 14]. After surface functionalisation and ion exchange to introduce the V precursor, the materials were calcined at 823 K in air. Calcination results in decomposition of both precursor and functionalisation agent. Preparation, functionalisation, and thermal treatment also have been described in Ref [9, 10, 14]. The authors stated that residuals of the functionalisation agent were no longer detectable (i.e. IR, Raman, UV-Vis, and XPS) in the material obtained. A brief summary of the N_2 physisorption analysis described in Ref [10] is given in Table 1.

Here, we have performed a detailed XAFS investigation of samples with different V loadings in the hydrated and dehydrated state. In particular, we wanted to analyze the contribution of higher scattering shells to the XAFS signal and possibly reveal the presence of V nearest neighbors in the local structure of vanadium oxide species supported on SiO_2. A detailed XAFS analysis of higher shells in the FT($\chi(k)*k^3$) has been largely neglected in the corresponding literature.

During thermal treatment of as-prepared hydrated V_xO_y/SBA-15 in oxygen (20% in He) a loss of water and a distinct change in structure were observed. After thermal treatment dehydrated V_xO_y/SBA-15 was cooled to 293 K in oxygen in He without exposure to air or water. No changes in XAFS spectra were observed during cooling. The EXAFS $\chi(k)*k^3$ of dehydrated V_xO_y/SBA-15 with different V loadings are depicted in Figure 1. The usable spectral ranged extended from 2.7 Å through 11.0 Å. The V K edge XANES spectra and the FT($\chi(k)*k^3$) of dehydrated V_xO_y/SBA-15 samples measured at 293 K are shown in Figure 2. FT($\chi(k)*k^3$) are not phase shift corrected. Thus, the distances in the FT($\chi(k)*k^3$) are shifted by ~0.4 Å to lower values compared to crystallographic distances. Compared to vanadium oxide references, the overall XANES region of dehydrated V_xO_y/SBA-15 resembled best those of NH_4VO_3, $Mg_2V_2O_7$, and Na_3VO_4 (Figure 3(a)). In the local structure of these references vanadium centers are tetrahedrally coordinated by four oxygen atoms. Compared to the XANES spectrum of dehydrated V_xO_y/SBA-15, NH_4VO_3 and $Mg_2V_2O_7$ exhibit very similar pre-edge peak heights in their XANES spectra (i.e. 0.65). Conversely, the pre-edge peaks in the XANES of Na_3VO_4 and $Mg_3V_2O_8$ are much higher than that of dehydrated V_xO_y/SBA-15 (Figure 3(a)).

In Figure 3(b) the $FT(\chi(k)*k^3)$ of NH_4VO_3, $Mg_2V_2O_7$, and Na_3VO_4 are compared to that of dehydrated V_xO_y/SBA-15. The first V-O peak in the $FT(\chi(k)*k^3)$ at ~1.4 Å (not phase shift corrected) for all references shown corresponds to a VO_4 tetrahedron in the respective structures. Apparently, the spectra of NH_4VO_3 and $Mg_2V_2O_7$ most closely resemble that of dehydrated V_xO_y/SBA-15. In the $FT(\chi(k)*k^3)$ of Na_3VO_4 and $Mg_3V_2O_8$ the first V-O peak is significantly higher than in the $FT(\chi(k)*k^3)$ of dehydrated V_xO_y/SBA-15. In contrast to the $FT(\chi (k)*k^3)$ of $Mg_3V_2O_8$ which exhibits a significant amplitude at distances above 2 Å, the $FT(\chi (k)*k^3)$ of Na_3VO_4, NH_4VO_3, $Mg_2V_2O_7$, and dehydrated V_xO_y/SBA-15 show little amplitude at higher distances. Moreover, looking at the differences between the $FT(\chi (k)*k^3)$ of Na_3VO_4, NH_4VO_3, and $Mg_2V_2O_7$, the latter appears to yield the best agreement with that of dehydrated V_xO_y/SBA-15. In all reference the low amplitude of the $FT(\chi (k)*k^3)$ at $R > 2$ Å is characteristic of the local structure around the V centers. It is not caused by an increased amount of disorder. In total, based on comparing the XANES and $FT(\chi (k)*k^3)$ of dehydrated V_xO_y/SBA-15 to those of potential references, NH_4VO_3 and $Mg_2V_2O_7$ have been identified as suitable references to serve as model systems for a more detailed structural analysis.

Table 1. N_2 physisorption analysis of supported vanadium oxide samples.

	Vanadium loading on SBA-15			S_{BET}	d_p	V_P
	(wt %)	V atoms/nm^2	(mmol/g)	(m^2/g)	(nm)	(mL/g)
SBA-15	-	-	-	897	7.0	1.1
2.7 wt % V/SBA-15	2.7	0.7	0.53	445	6.7	0.5
5.4 wt % V/SBA-15	5.4	1.4	1.05	440	6.6	0.5
10.8 wt % V/SBA-15	10.8	4.7	2.12	273	5.5	0.3

Vanadium loading, surface area (S_{BET}), pore diameter (d_p), and pore volume (V_p) of SBA-15 and vanadium oxides supported on SBA-15. Details have been presented in Ref [10].

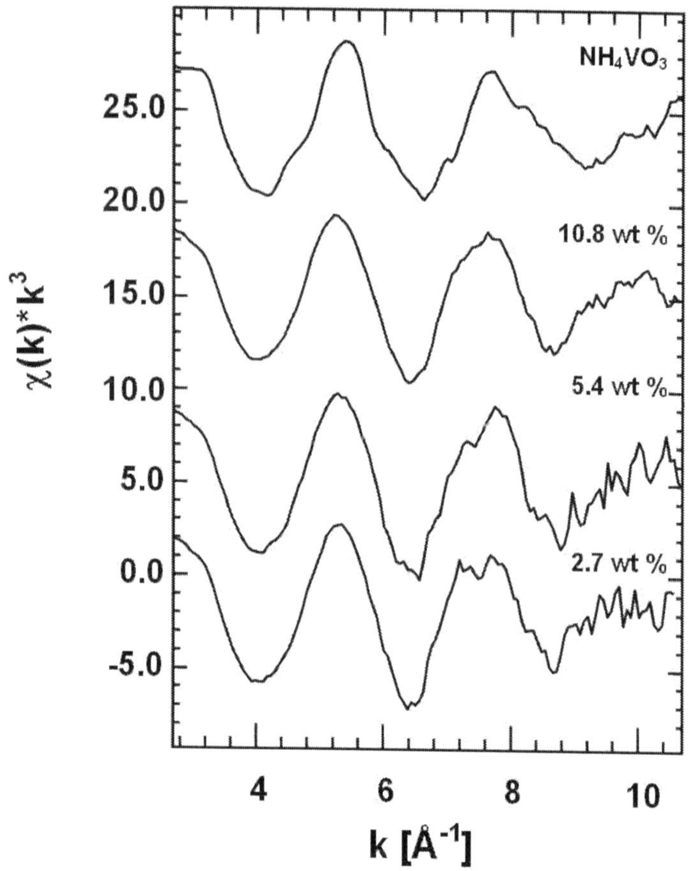

Figure 1. V K edge $\chi(k)$ of dehydrated V_xO_y/SBA-15 with different vanadium loadings (2.7 wt %, 5.4 wt %, and 10.8 wt %) and reference NH_4VO_3.

Before we discuss the details of analyzing the XAFS data of dehydrated V_xO_y/SBA-15, a suitable analysis procedure for the higher V-V contributions in the XAFS spectra of references NH_4VO_3 and $Mg_2V_2O_7$ was sought. As an example and to reduce the number of tables here, the application of confidence limits and F parameter to distinguish analysis fitting procedures is described for two refinements of a suitable model structure to the experimental FT(χ (k)*k^3)of dehydrated V_xO_y/SBA-15. The model structure consisted of a tetrahedral coordination of the V center by four oxygen atoms at ~1.7 Å (NH_4VO_3), two vanadium atoms at distances at ~3.4 Å (NH_4VO_3) and 3.6 Å ($Mg_2V_2O_7$), one oxygen atom at ~2.9 Å ($Mg_2V_2O_7$), and one Si atom at ~2.8 Å (Table 2). Experimental FT(χ(k)*k^3) of $Mg_2V_2O_7$ and NH_4VO_3 and the corresponding XAFS refinements are shown in Figure 4. Deviations between the theoretical and experimental spectrum of $Mg_2V_2O_7$ in the range from 2 - 4 Å are caused by the number of Mg neighbors and nearly linear multiple-scattering paths in $Mg_2V_2O_7$ that contribute in this range. These are not sufficiently accounted for by the simplified refinement procedure. The results of the XAFS refinement for dehydrated V_xO_y/SBA-15, NH_4VO_3, and $Mg_2V_2O_7$ are summarized in Table 2. Apparently, the distorted VO_4 tetraeder in NH_4VO_3 required two different V-O distances to be included in the refinement, while $Mg_2V_2O_7$ and dehydrated V_xO_y/SBA-15 exhibited a similar single V-O distance. It seems that the distortion in the VO_4 units of crystalline reference $Mg_2V_2O_7$ could not be resolved by the XAFS analysis procedure employed. Accordingly, a lower σ^2 was obtained (0.0012 Å^2) for NH_4VO_3 compared to those of dehydrated V_xO_y/SBA-15 and $Mg_2V_2O_7$ (0.0075 Å^2 and 0.0059 Å^2, respectively). Contrarily, a single V-V distance at 3.47 Å (CN = 2) sufficed for NH_4VO_3 (consisting of chains of VO_4 units), while two V-V distances had to be included for $Mg_2V_2O_7$ (consisting of adjacent V_2O_7 units) and dehydrated V_xO_y/SBA-15. In all three cases, a similar σ^2 parameter for the V-V contributions of about 0.014 Å^2 was obtained.

The corresponding confidence limits and significance parameters F are given in Table 3. In fitting procedure #1 two V-O distances in the first V-O shell were allowed to vary independently (both with a CN of 2 and the same σ^2). Moreover, E_0 was also allowed to vary in fitting procedure #1. Because of N_{ind} = 18 and N_{free} = 11 refinement procedure #1 would be taken as reliable according to the Nyquist criteria. However, confidence limits of the first V-O distance of ± 0.1 Å and an F parameter of 0.7 for both distance and σ^2 were obtained with procedure #1. Moreover, E_0 exhibited a confidence limit of ± 9.2 and F = 0.9. Apparently, fitting procedure #1 already exceeds the number of meaningful parameters and yield ambiguous structural parameters. On the one hand, the reduced amplitude of the FT(χ(k)*k^3) of dehydrated V_xO_y/SBA-15 compared to that of references consisting of undistorted VO_4 units (Figure 3) suggested the presence of more than one V-O distance, including a short "vanadyl" V = O distance. On the other hand, however, the resolution in the experimental FT(χ(k)*k^3) and the available degree of freedom did not permit refining more than one V-O distance in the procedure used.

Therefore, the fitting procedure was modified. In the following, E_0 was kept invariant in the refinement and only one V-O distance at ~1.75 Å was used. In contrast to procedure #1, procedure #2 yielded reasonable confidence limits and acceptable F parameters.

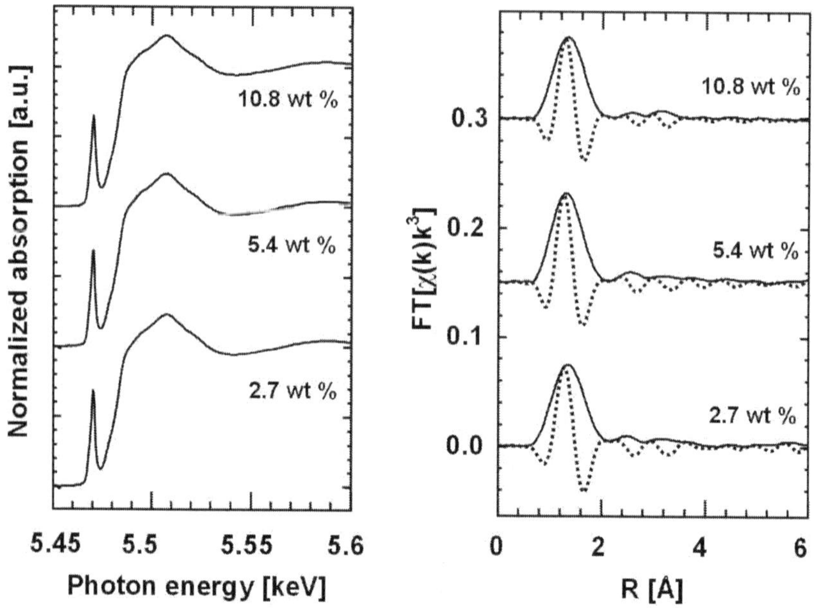

Figure 2. V K edge XANES spectra (a) and FT(χ(k)*k ^{3}) (b) of dehydrated V_xO_y/SBA-15 samples with different vanadium loadings (2.7 wt %, 5.4 wt %, and 10.8 wt %).

The σ^2 parameter of the V-O contribution at 2.9 Å exhibited a rather high confidence limit and F = 0.7. Apparently, both V-O and V-Si neighbors in the distance range from 2.5 Å to 2.9 Å are required for a good refinement of the model structure to the experimental data. This is indicated by the confidence limits and F parameters calculated for the corresponding distances (Table 3). Nevertheless, the high σ^2 obtained for the V-Si contribution and the rather low σ^2 obtained for the V-O at 2.9 Å indicate a certain ambiguity of the corresponding fitting results. The reason may be a considerable static disorder and, thus, a broadened V-Si distance distribution. Hence, calculating and evaluating confidence limits and F tests permitted to arrive at a meaningful and reliable fitting procedure. In that, the approach employed appears to be superior to only calculating the Nyquist criteria. In total, procedure #2 worked very well for XAFS data analysis of dehydrated V_xO_y/SBA-15 and $Mg_2V_2O_7$. In contrast, the local structure around V centers in NH_4VO_3 was best described by assuming two different V-O distances in the first coordination shell and only one V-V distance at 3.47 Å (CN = 2) (Table 2). A V-O distance at 2.8 Å was found to be insignificant.

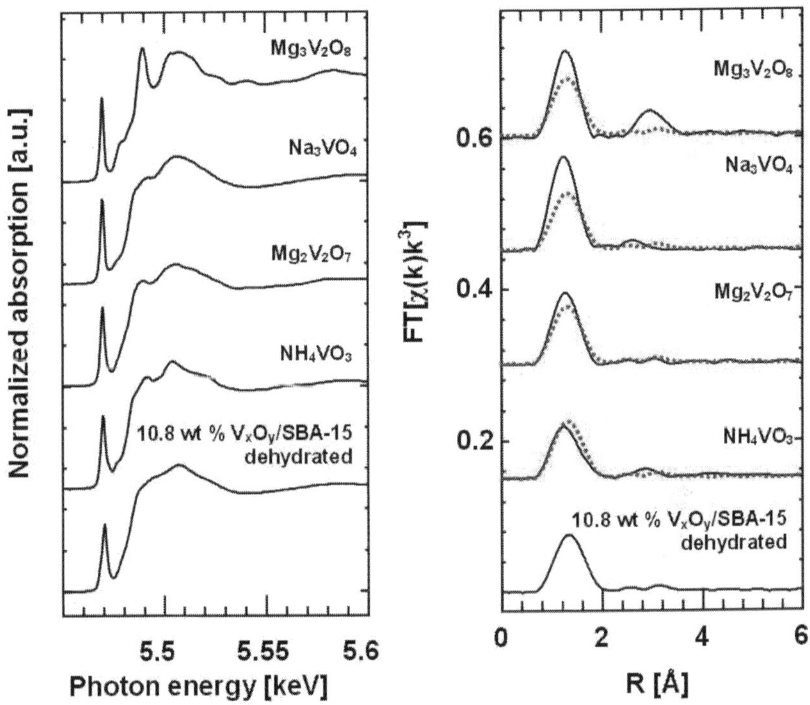

Figure 3. V K edge XANES spectra (a) and FT(χ(k)*k ^{3}) (b) of dehydrated V_xO_y/SBA-15 (10.8 wt %) (dashed on right side) compared to those of various references (i.e. NH_4VO_3, $Mg_2V_2O_7$, Na_3VO_4, and $Mg_3V_2O_8$).

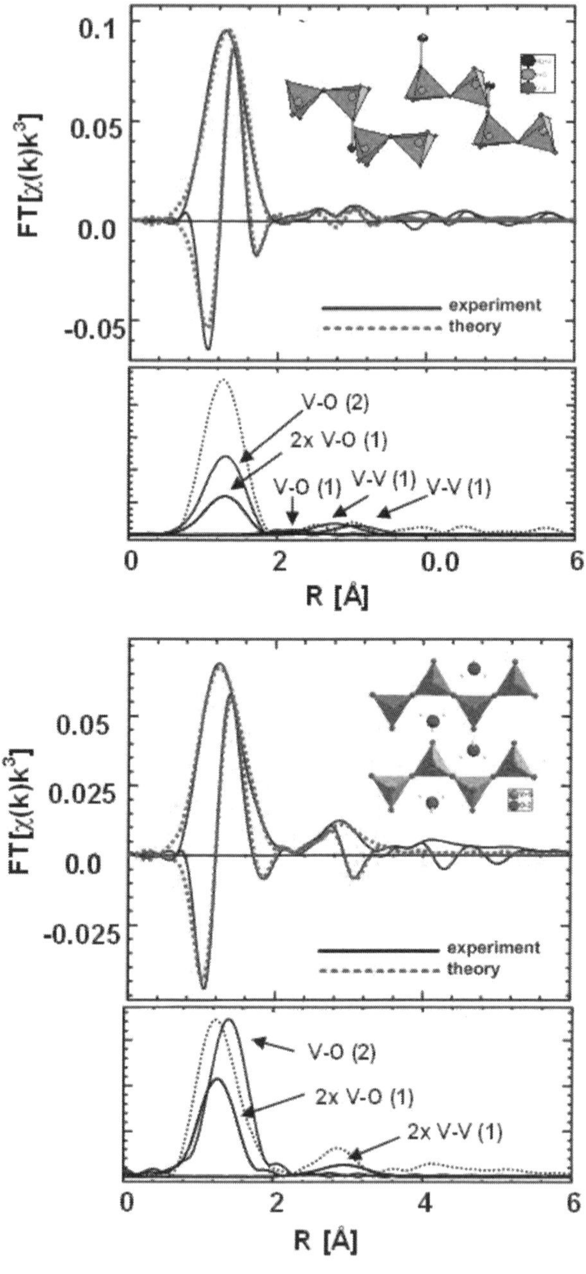

Figure 4 Experimental (solid) V K edge FT($\chi(k)*k^3$) of NH_4VO_3, $Mg_2V_2O_7$ reference together with a theoretical XAFS function (fitting results are given in Table 2). Also shown are the Fourier transformed $\chi(k)*k^3$ of the individual scattering paths together with corresponding coordination number in brackets. brackets.

Table 2 . EXAFS refinement results obtained for experimental $FT(\chi(k)*k^3)$ of dehydrated V_xO_y/SBA-15, NH_4VO_3, and $Mg_2V_2O_7$.

	Model		V_xO_y-SBA-15		$Mg_2V_2O_7$		NH_4VO_3	
Type	N	R [Å]	R [Å]	σ^2 [Å^2]	R [Å]	σ^2 [Å^2]	R [Å]	σ^2 [Å^2]
V-O	1	1.63	1.78	0.0075	1.74	0.0059	1.68	0.0012
V-O	1	1.70	1.78_C	0.0075_C	1.74_C	0.0059_C	1.68_C	0.0012_C
V-O	2	1.76	1.78_C	0.0075_C	1.74_C	0.0059_C	1.84	0.0012_C
V-O	1	2.87	2.89	0.0017	2.77	0.0165	-	-
V-V	1	3.36	3.30	0.0135	3.28	0.0134	3.47	0.0147
V-V	1	3.62	3.62	0.0135_C	3.56	0.0134_C	3.47_C	0.0147_C
V-Si	1	2.80	2.54	0.0121	-	-	-	-

Type and number (N) of atoms at distance R from the absorbing V atom in a model system assuming an ordered arrangement of V_2O_7 units (Figure 7) compared to experimental distances and XAFS disorder parameters (σ^2). Parameters were obtained from the refinement of this model structure to the experimental V K edge XAFS $FT(\chi(k)*k^3)$ of dehydrated V_xO_y/SBA-15 (10.8 wt %), $Mg_2V_2O_7$, and NH_4VO_3 (k range from 2.7-11.0 Å^{-1}, R range 0.8-4.0 Å, N_{ind} = 18, E_0 = 0 eV in all cases, fit residual 3.6 (dehydrated V_xO_y/SBA-15) (N_{free} = 9), 11.2 ($Mg_2V_2O_7$) (N_{free} = 7), 8.9 (NH_4VO_3) (N_{free} = 5)) (Subscript C indicates parameters that were correlated in the refinement). Confidence limits and significance of fitting parameters are given in Table 3.

2.2 Local structure of dehydrated V_xO_y/SBA-15 - XAFS refinement of "VO_4" based model structures

After having identified two suitable references as model structures for XAFS refinements to the experimental $FT(\chi(k)*k^3)$ of dehydrated V_xO_y/SBA-15 (Figure 3(b)), the XAFS analysis approach chosen shall be described in more detail. In addition to using confidence limits and F tests as introduced above, the suitable XAFS fitting procedure was developed stepwise as outlined in the following.

Table 3 . Evaluation of EXAFS refinement of dehydrated V_xO_y/SBA-15.

Type R [Å] σ^2 [Å^2]		Procedure #1			Procedure #2		
	N	Z	± z	F	Z	± z	F
R(V-O)	4(2)	1.81	0.11	0.7	1.78	0.005	0
σ^2(V-O)	4	0.0066	0.0047	0.7	0.0075	0.0004	0
R(V-O)	-(2)	1.75	0.04	0.4	-	-	
R(V-O)	1	2.89	0.01	0.5	2.90	0.011	0
σ^2(V-O)	1	0.0014	0.0017	0.9	0.0017	0.0018	0.7
R(V-V)	1	3.29	0.017	0	3.29	0.016	0
σ^2(V-V)	2	0.0135	0.0203	0.7	0.0135	0.00035	0.3
R(V-V)	1	3.61	0.019	0	3.62	0.024	0
R(V-Si)	1	2.54	0.01	0	2.54	0.011	0
σ^2(V-Si)	1	0.0115	0.0011	0.3	0.0121	0.0011	0
E_0	-	-0.9	-0.3	0.8	-	-	-

V K edge XAFS parameters (Z for distances R and disorder parameter σ^2) obtained from two different procedures of fitting a model structure (i.e. "ordered V_2O_7 dimers" on SiO_2 support) to the experimental XAFS FT(χ(k)*k^3) of dehydrated V_xO_y/SBA-15 (10.8 wt %) (details of fit given in Table 2) together with confidence limits (± z, referring to 95% of fit residual) and significance parameters F (details given in text). Fit residual 3.1 for Procedure #1 and 3.6 for Procedure #2.

First, we started with an often repeated assumption from the literature. DR-UV-Vis or Raman measurements revealed that dehydration of V_xO_y/SBA-15 resulted in a characteristic change from a distorted square pyramidal to a distorted tetrahedral coordination [10, 13, 16]. The local structure of vanadium oxide species supported on SiO_2 was assumed to correspond to isolated VO_4 units. Hence, in a first *tetrahedron approach* the theoretical XAFS function of a VO_4 tetrahedron consisting of two slightly different V-O distances was refined to the FT(χ(k)*k^3) of dehydrated V_xO_y/SBA-15 (Figure 5, top) together with the Fourier transformed χ(k)*k^3 of the individual scattering paths). Because of the similar height of the pre-edge peak in the XANES (Figure 3(a)) and the first V-O peak in the FT($\chi(k)*k^3$), phases and amplitudes employed in the refinement were calculated using the model structure of NH_4VO_3 (ICSD 1487 [26]) and $Mg_2V_2O_7$ (ISCD 2321 [27]). Figure 5, top shows a good agreement between theoretical and experimental FT($\chi(k)*k^3$) of dehydrated V_xO_y/SBA-15 for the first V-O peak below 2 Å. Naturally, the amplitude between 2 Å and 4 Å in the FT($\chi(k)*k^3$) could not be accounted for. Hence, a structural model assuming only isolated VO_4 species cannot adequately describe the local structure around the V centers in dehydrated V_xO_y/SBA-15.

Therefore, we assumed that higher coordination shells around the vanadium centers significantly contribute to the FT($\chi(k)*k^3$) of dehydrated V_xO_y/SBA-15. These shells have to be included in the refinement. A seemingly expected contribution may arise from silicon backscatterers in the SiO_2 support at distances of less than 3.0 Å. This has been previously proposed by Keller et al. [20]. Thus, in extension of the *tetrahedron approach* a silicon atom at a V-Si distance of 2.8 Å was included in the theoretical model. In the corresponding "O_3V-O-Si" unit a Si-O distance of 1.62 Å is assumed (inset in Figure 5, bottom)), as it is found in various silicates. The result of the XAFS refinement of the "O_3V-O-Si" model to the FT($\chi(k)*k^3$) of dehydrated V_xO_y/SBA-15 is depicted in Figure 5, bottom). The additional Si backscatterer resulted in a better agreement between theoretical and experimental FT($\chi(k)*k^3$) at distances of about 2.4 Å (not phase shift corrected). The resulting V-Si distance amounted to 2.54 Å, comparable to the distance obtained by Keller et al. (2.61 Å [20]). However, it can be easily seen from Figure 5, bottom) that the amplitude in the FT($\chi(k)*k^3$) of dehydrated V_xO_y/SBA-15 between 2.4 and 4.0 Å is still not accounted for.

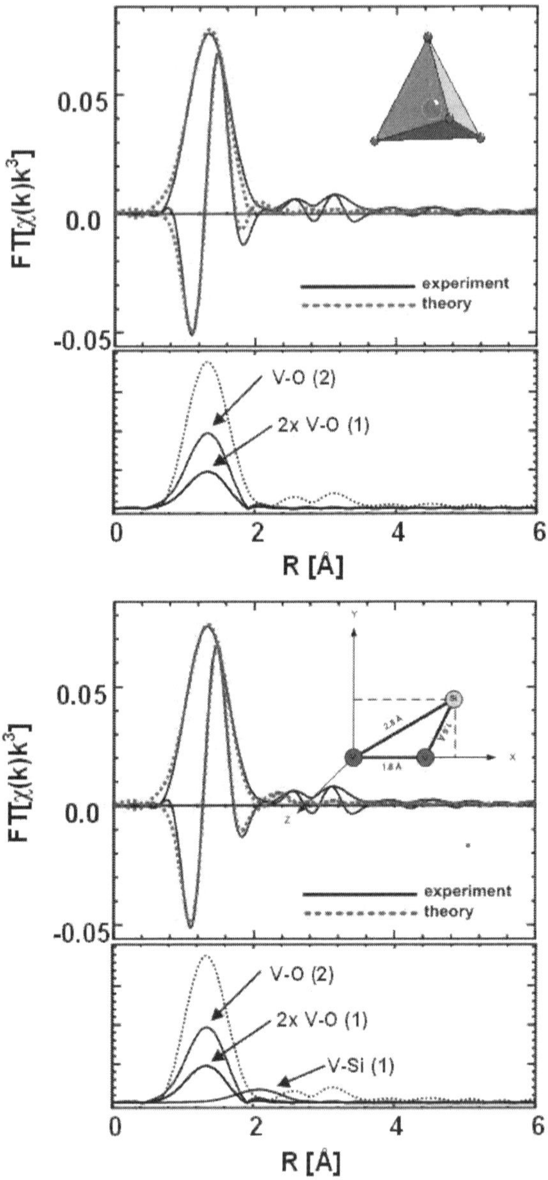

Figure 5. Experimental (solid) V K edge FT($\chi(k)*k^3$) of dehydrated V_xO_y/SBA-15 (10.8 wt %) together with theoretical XAFS functions (top: "isolated VO_4" model, bottom: addition of V-Si path to "isolated VO_4" model). Insets show the VO_4 tetrahedron (top) and a schematic representation of the V-Si path employed (bottom). Also shown are the Fourier transformed $\chi(k)*k^3$ of the individual scattering paths together with corresponding coordination number in.

2.3 Local structure of dehydrated V_xO_y/SBA-15 - XAFS refinement of "V_2O_7" based model structures

Figure 5 shows that an "isolated VO_4" model did not properly describe the local structure between 2 Å and 4 Å around vanadium centers in dehydrated V_xO_y/SBA-15. Hence, we assumed that at least "V_2O_7 dimers" would be needed to achieve a good agreement between theoretical and experimental XAFS FT(χ(k)*k^3). V_2O_7 units are present in the structures of the references NH_4VO_3 and $Mg_2V_2O_7$ whose spectra resembled best the XANES and EXAFS spectra of dehydrated V_xO_y/SBA-15 (Figure 3). Therefore, a V-V scattering path at 3.4 Å was included in the model used in the XAFS refinement. This distance corresponds to the shortest V-V distance between two corner-sharing VO_4 tetrahedrons in "V_2O_7 dimers" of NH_4VO_3 and $Mg_2V_2O_7$. The result of the corresponding XAFS refinement is shown in. Apparently, a structural model based on isolated V_2O_7 dimers was equally unsuited to describe the local structure around V centers in dehydrated V_xO_y/SBA-15. The agreement between theoretical and experimental FT(χ(k)*k^3) in the range from 2 to 4 Å is still not sufficient (Figure 6). Also, adding a V-Si distance to this "isolated V_2O_7 dimer model" only resulted in a minor improvement of the refinement.

Figure 5 and Figure 6 clearly show that neither an "isolated VO_4" model nor an "isolated V_2O_7" model properly describe the local structure of the majority of V centers in dehydrated V_xO_y/SBA-15. Hence, in the next step an ordered arrangement of neighboring V_2O_7 units was assumed. Because of their similar XANES and EXAFS spectra, we again referred to NH_4VO_3 and $Mg_2V_2O_7$ as references. V_2O_7 units form chains in NH_4VO_3 with one V-V distance. Conversely, V_2O_7 units are neighboring but more separated in $Mg_2V_2O_7$ resulting in two distinct V-V distances (ICSD 2321 [27]). Accordingly, two additional scattering paths were added to the previous "isolated V_2O_7" model. These two paths correspond to V-O (2.8 Å) and V-V (3.6 Å) distances between two neighboring V_2O_7 units in the structure of $Mg_2V_2O_7$. The result of the corresponding XAFS refinement to the FT(χ(k)*k^3) of dehydrated V_xO_y/SBA-15 is shown in Figure 7 together with the various V-O and V-V distances used. Apparently, assuming neighboring V_2O_7 units in an ordered arrangement supported on SBA-15 yielded a good agreement between theoretical and experimental FT(χ(k)*k^3) of dehydrated V_xO_y/SBA-15 over the extended R range from 1 Å to 4 Å. The structural and fitting parameters obtained from the XAFS refinement to the experimental FT(χ(k)*k^3) of dehydrated V_xO_y/SBA-15 and $Mg_2V_2O_7$ are given in Table 2. The similar V-O distances, V-V distances, and σ^2 parameters of dehydrated V_xO_y/SBA-15 and $Mg_2V_2O_7$ corroborate our choice of model system to describe the local structure around V centers dehydrated V_xO_y/SBA-15.

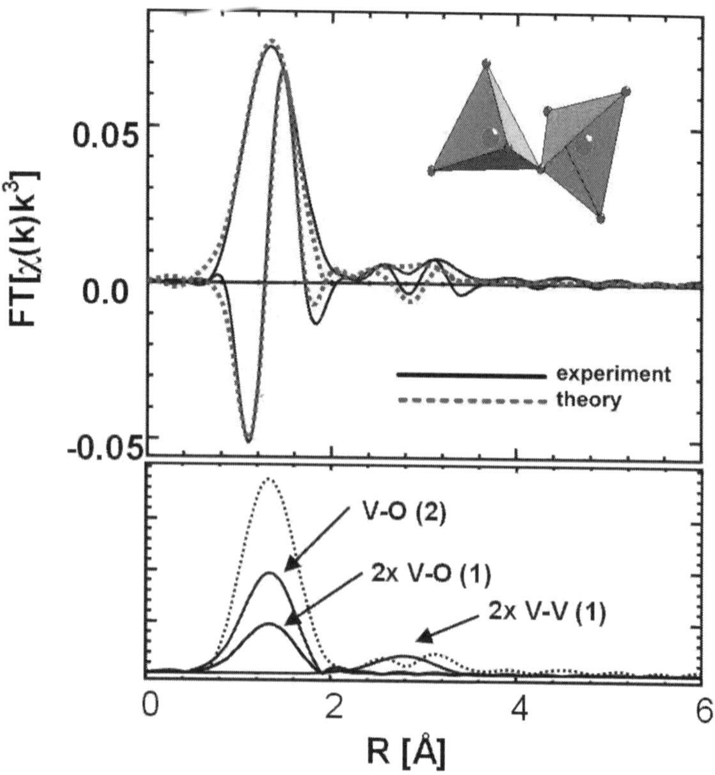

Figure 6. Experimental (solid) V K edge FT(χ(k)*k^3) of dehydrated V_xO_y/SBA-15 (10.8 wt %) together with a theoretical XAFS function (i.e. "isolated V_2O_7" model). Inset shows the V_2O_7 dimer. Also shown are the Fourier transformed χ(k)*k^3 of the individual scattering paths together with corresponding coordination number in brackets.

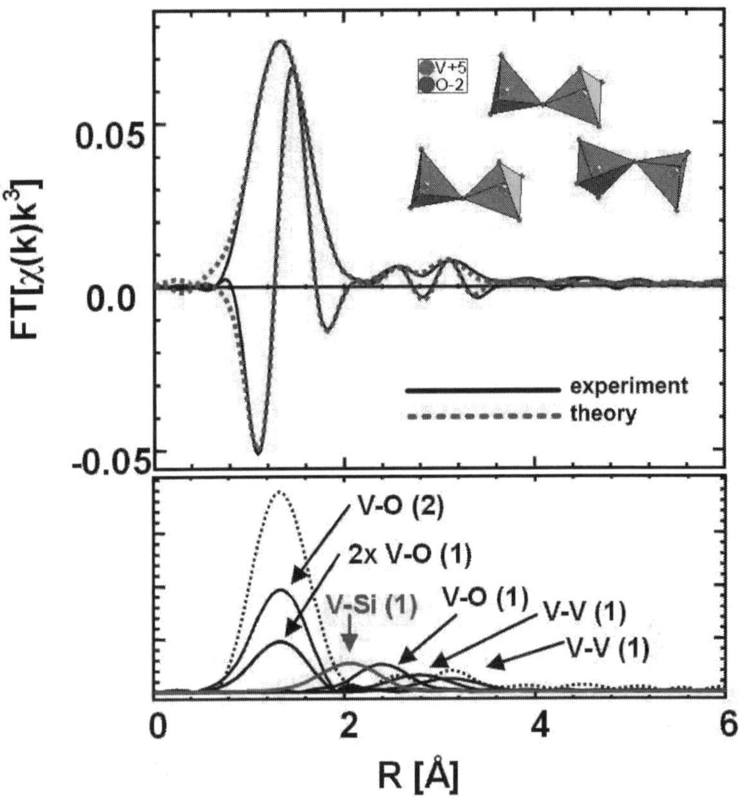

Figure 7. Experimental (solid) V K edge FT(χ(k)*k^3) of dehydrated V_xO_y/SBA-15 (10.8 wt %) together with a theoretical XAFS function (i.e. "ordered arrangement of V_2O_7" model). Fitting results are given in Table 2. Inset shows a schematic representation of arrangement of V_2O_7 units in $Mg_2V_2O_7$. Also shown are the Fourier transformed χ(k)*k^3 of the individual scattering paths together with corresponding coordination number in brackets.

2.4 Schematic structural representation of dehydrated V_xO_y/SBA-15

A schematic structural representation of the ordered arrangement of V_2O_7 units in dehydrated V_xO_y/SBA-15 is depicted in Figure 8. In contrast to previous results on low loaded (< 1 V/nm^2) V_xO_y/SiO_2 samples [17, 21] we conclude that isolated VO_4 units are not the major vanadium oxide species present on the dehydrated V_xO_y/SBA-15 samples studied here. From the different loadings studied, only the 2.7 wt % V_xO_y/SBA-15 sample possessed a vanadium content of less than 1 V/nm^2. The three dehydrated V_xO_y/SBA-15 samples exhibited only minor differences in their XANES spectra (Figure 2), FT(χ(k)*k^3) (Figure 9), and XAFS fitting results (Table 4). Hence, independent of the V loading in the range 2.7 - 10.8 wt % the local structure of the majority of V centers in dehydrated V_xO_y/SBA-15 is described best by an ordered arrangement of neighboring V_2O_7 units (Table 4, Figure 9). Decomposition of the decavanadate precursor during calcinations of the as-prepared materials will most likely result in the formation of dehydrated V_xO_y/SBA-15. Any description of the formation mechanism of ordered V_2O_7 units on the surface of SiO_2, however, is beyond the scope of this work. Exposure of the calcined material to ambient conditions apparently results in re-hydration and formation of the hydrated V_xO_y species supported on SBA-15. This transformation already suggests a reversible hydration-re-hydration behavior of vanadium oxide species supported on SBA-15 which should be the subject of further studies.

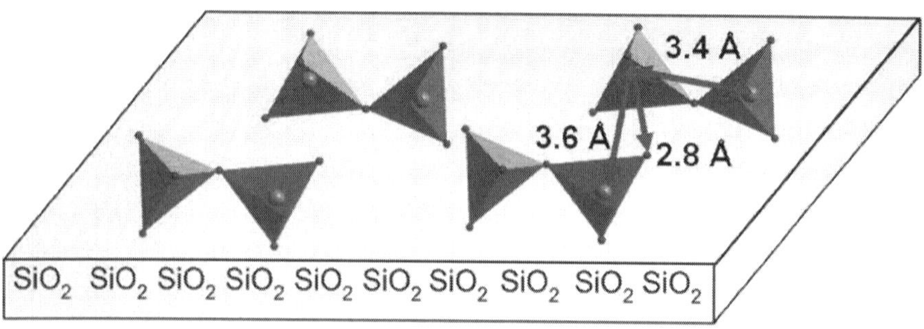

Figure 8. Schematic structural representation of dehydrated V_xO_y/SBA-15. The most prominent distances employed in the XAFS refinement procedure are indicated.

The presence of non-monomeric V_xO_y species in dehydrated V_xO_y/SBA-15 samples was also recently concluded based on NEXAFS studies combined with theoretical calculations [28]. Eventually, structural evolution and catalytic activity of dehydrated V_xO_y/SBA-15 were studied by combined in situ XAS-MS under selective propene oxidation reaction condition. Onset of catalytic activity was detected at about 573 K. The in situ XAS data measured indicated, that the characteristic ordered arrangement of V_2O_7 dimers in the local structure of dehydrated V_xO_y/SBA-15 persisted under catalytic reaction conditions. A more detailed analysis of structure activity correlations of V_xO_y/SBA-15 under selective oxidation reaction conditions will be presented elsewhere.

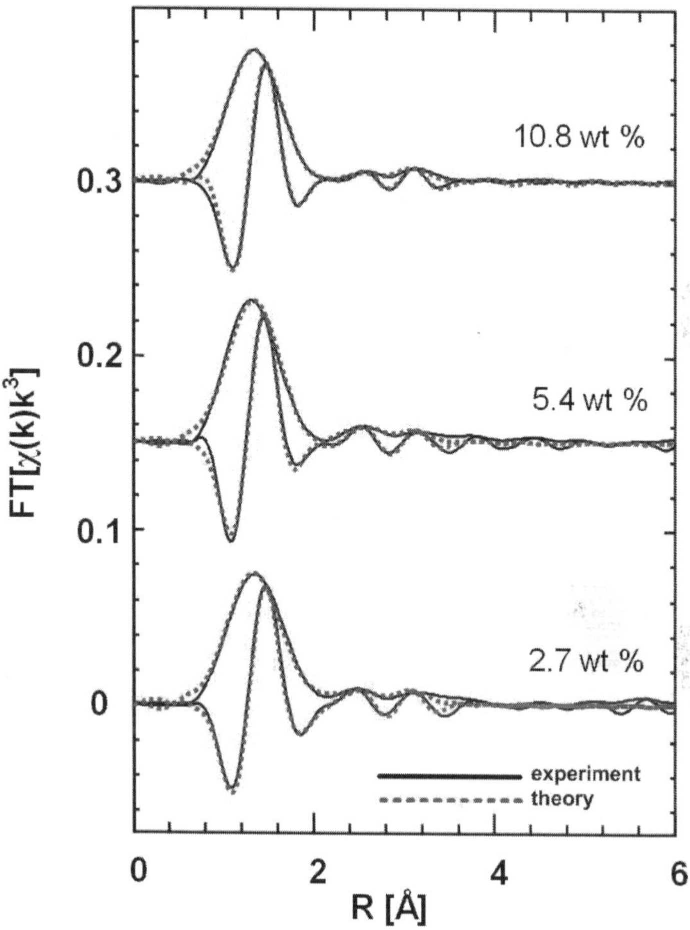

Figure 9. Experimental (solid) V K edge FT($\chi(k)*k^3$) of dehydrated V_xO_y/SBA-15 (2.7 wt %, 5.4 wt %, and 10.8 wt %) together with a theoretical XAFS function (structural model depicted in Figure 7). Fitting results are given in Table 4.

Table 4. EXAFS refinement results obtained for different V loadings on V_xO_y/SBA-15 in the dehydrated state.

Type	Model		10.8 wt %		5.4 wt %		2.7 wt %	
	N	R [Å]	R [Å]	σ^2[Å^2]	R [Å]	σ^2[Å^2]	R [Å]	σ^2[Å^2]
V-O	1	1.63	1.78	0.0075	1.77	0.0069	1.78	0.0075
V-O	1	1.70	1.78	0.0075	1.77	0.0069	1.78	0.0075
V-O	2	1.76	1.78	0.0075	1.77	0.0069	1.78	0.0075
V-V	1	3.36	3.30	0.0135	3.33	0.0114	3.37	0.0113
V-O	1	2.87	2.90	0.0017	2.87	0.0022	2.86	0.0015
V-V	1	3.62	3.62	0.0135	3.62	0.0114	3.65	0.0113
V-Si	1	2.80	2.54	0.0121	2.53	0.0139	2.54	0.0081

Type and number (N) of atoms at distance R from the absorbing V atom in a model system assuming an ordered arrangement of V_2O_7 units (Figure 7) compared to experimental distances and XAFS disorder parameter (σ^2). Parameters were obtained from the refinement of this model structure to the experimental V K edge XAFS FT(χ(k)*k^3) of dehydrated V_xO_y/SBA-15 with different V loadings (i.e. 10.8 wt %, 5.4 wt %, 2.7 wt %) (Figure 9) (k range from 2.7-11.0 Å^{-1}, R range 0.8-4.0 Å, N_{ind} = 18, N_{free} = 9, E_0 = 0 eV in all cases, fit residual 3.6 (10.8 wt %), 7.0 (5.4 wt %), 7.8 (2.7 wt %)) (Subscript C indicates parameters that were correlated in the refinement). Confidence limits and significance of fitting parameters correspond to those given in for the 10.8 wt % sample.

Oxygen and silicon atoms of the SiO_2 support are not depicted in the schematic representation depicted in Figure 8. In particular Si atoms in the topmost layer of SiO_2 belong to the second coordination sphere of the V centers. Previous reports have indicated that V-Si distances may contribute to the experimental $FT(\chi(k)*k^3)$ of dehydrated V_xO_y/SBA-15 [19]. Therefore, a single V-Si scattering path was included in the refinement of the "neighboring V_2O_7" model described above (Figure 7). The structural parameters and refinement details are given in Table 2 and Table 4. Comparing fit residuals, confidence limits, and F parameters a significant improvement was visible. Apparently, both local structure in V_xO_y species and interaction with SiO_2 support are required to describe the $FT(\chi(k)*k^3)$ of dehydrated V_xO_y/SBA-15 samples.

2.5 Limitations of XAFS analysis of dehydrated V_xO_y/SBA-15

Eventually, the limitations of the XAFS analysis of dehydrated V_xO_y/SBA-15 presented here should be discussed. XAFS is not a very sensitive technique with respect to distinguishing and identifying additional minority species. Experimental XAFS spectra are clearly dominated by the signal of the majority phase. Hence, the presence of minority vanadium oxide species in dehydrated V_xO_y/SBA-15 with concentrations of less than ~5% cannot be excluded. Only if the contribution of additional phases amounts to more than ~5-10%, distortion of the $FT(\chi(k)*k^3)$ and deviation from the model structure assumed will be detectable. This holds in particular, if these minority species happen to be less ordered than the majority phase.

Moreover, XAFS is an averaging technique. Certainly, higher shells should be properly taken into account and various references should be measured for comparison. Even then, however, it may remain difficult to unambiguously distinguish between mixtures of various species or structures. Hence, alternative scenarios with different vanadium oxide species need to be considered and discussed. An equal mixture of isolated VO_4 and neighboring V_2O_7 units, for instance, may exhibit a XAFS $FT(\chi(k)*k^3)$ similar to that presented here. For two reasons this assumption is not very likely. First, the V-O distances in the first "VO_4" shell of the two species would have to be the same. Otherwise a strong reduction in amplitude of the first V-O peak in the FT(χ(k)*k^3) caused by destructive interference would be discernible. Second, isolated VO_4 would not contribute to the FT(χ(k)*k^3) in the range from 2 Å to 4 Å. Thus, reduction in amplitude and much higher σ^2 parameters compared to reference $Mg_2V_2O_7$ would be detectable. Both are not observed in the EXAFS analysis of dehydrated V_xO_y/SBA-15 presented here (Figure 3(b), Table 2). A similar statement holds for isolated V_2O_7 units or a less ordered arrangement of neighboring V_2O_7 units. Both would result in a reduction in FT(χ(k)*k^3) amplitude because of missing contribution in the 2-4 Å range or destructive interference caused by a broadened distribution of distances, respectively. Moreover, for isolated or less ordered V_2O_7 units a single V-V distance would suffice to describe the experimental XAFS spectrum. This was also not observed in the XAFS analysis procedure employed. Eventually, higher V-Si distances may have to be considered in addition to a V-Si distance of ~2.5 Å (Table 2). However, a significant contribution of V-Si distances at more 3.0 Å range in the FT(χ(k)*k^3) would require a highly ordered arrangement of V_xO_y species on the SiO_2 support and a very narrow distance distribution. This seems to be unlikely.

In total, assuming a structural arrangement of vanadium centers in dehydrated V_xO_y/SBA-15 that has already been established for reference vanadium oxides (i.e. $Mg_2V_2O_7$) is simple and results in a good agreement with experimental data. More complex and artificially constructed arrangements of V_xO_y species supported on SiO_2 may be conceivable but appear to be less likely. Both XANES and EXAFS analysis corroborate a local structure around the majority of V centers in dehydrated V_xO_y/SBA-15 similar to the ordered arrangement of neighboring V_2O_7 dimers in the structure of $Mg_2V_2O_7$.

2.6 Local structure of hydrated V_xO_y/SBA-15 Comparison to V oxide references

The EXAFS χ(k)*k^3 of hydrated V_xO_y/SBA-15 (as-prepared) with different V loadings are depicted in Figure 10. The usable spectral ranged extended from 2.7 Å through 10.5 Å. The V K edge XANES spectra and the FT(χ(k)*k^3) of hydrated V_xO_y/SBA-15 are shown in Figure 11. The Fourier transformed χ(k)*k^3 and the V K near edge spectra of hydrated V_xO_y/SBA-15 are compared to those of vanadium oxide references in Figure 12. The XANES spectrum of hydrated V_xO_y/SBA-15 resembles that of MgV_2O_6, $[H_3N(CH_2)_4]_6V_{10}O_{28}$, and V_2O_5 (Figure 12(a)). In these vanadium oxide references vanadium centers exhibit a distorted octahedral or distorted square pyramidal coordination. The XANES spectra of hydrated V_xO_y/SBA-15, V_2O_5, MgV_2O_6, and $[H_3N(CH_2)_4]_6V_{10}O_{28}$ show a similar height of the pre-edge peak. Because the pre-edge peak height is determined by the coordination of the vanadium centers [29], hydrated V_xO_y/SBA-15 also appears to exhibit a distorted square pyramidal coordination of V centers. This has also been observed by Bell et al. [18] and others.

The range of potential model structures describing the local structure of hydrated V_xO_y/SBA-15 can be further narrowed when comparing the corresponding FT(χ(k)*k^3) (Figure 12(b)). Considering peak positions and relative peak heights in the FT(χ(k)*k^3), it appears that from the references available the FT(χ(k)*k^3) of V_2O_5 resembles that of hydrated V_xO_y/SBA-15. Because of the lower intensity in the FT(χ(k)*k^3) hydrated V_xO_y/SBA-15 may possess a more disordered structure compared to that of crystalline V_2O_5. This is in good agreement with a detailed comparison of the corresponding XANES spectra. On the one hand, the positions of the various peaks in the XANES of hydrated V_xO_y/SBA-15 are similar to that of V_2O_5 (Figure 12(a)). On the other hand, the lower peak intensities are also indicative of a disordered V_2O_5 like structure of the vanadium oxide species in hydrated V_xO_y/SBA-15.

2.7 Local structure of hydrated V_xO_y/SBA-15 XAFS refinement of "V_2O_5" based model structure

Comparison of the XANES and FT(χ(k)*k^3) of hydrated V_xO_y/SBA-15 to those of various references identified V_2O_5 as most suitable model structure for a detailed EXAFS analysis. Ammonium decavanadate decomposes to V_2O_5 during treatment in air at temperatures above 773 K. Calcination of the materials studied here will most likely result in formation of the dehydrated species as described above. Re-hydration upon exposure to ambient conditions resulted in vanadium oxide species supported on SBA-15 with a local structure similar to that of V_2O_5. A detailed discussion of the underlying formation mechanisms is beyond the scope of this work. Therefore, a theoretical XAFS function calculated on the basis of a V_2O_5 model structure (ICSD 60767 [30]) was refined to the experimental FT(χ(k)*k^3) of hydrated V_xO_y/SBA-15. Details of the XAFS refinement procedure and the structural parameters obtained are given in Table 5. In addition, the V_2O_5 model structure was refined to the FT(χ(k)*k^3) of V_2O_5 to validate the procedure chosen. Good agreement between the theoretical XAFS function of a V_2O_5 model structure and the FT(χ(k)*k^3) of hydrated V_xO_y/SBA-15 and V_2O_5 was obtained (Figure 13).

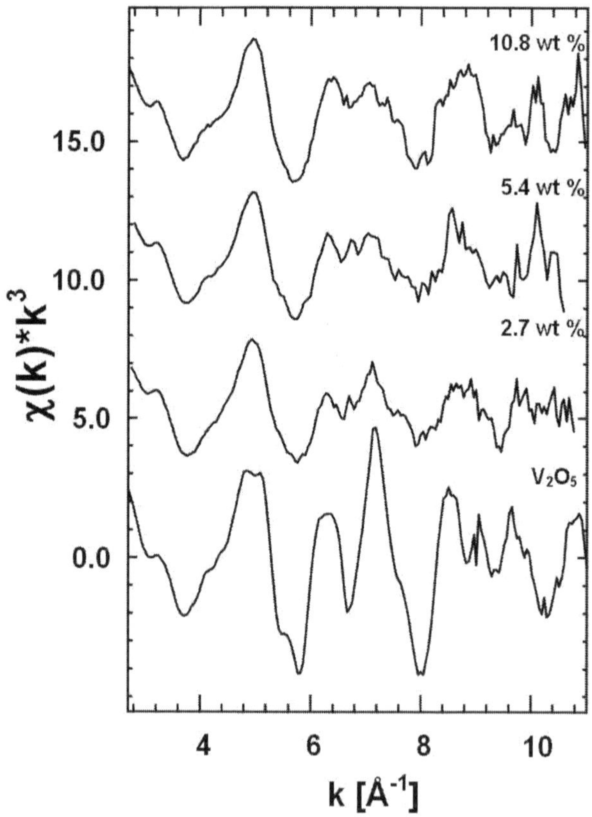

Figure 10. V K edge χ(k) of hydrated V_xO_y/SBA-15 samples with different vanadium loadings (2.7 wt %, 5.4 wt %, and 10.8 wt %) and reference V_2O_5.

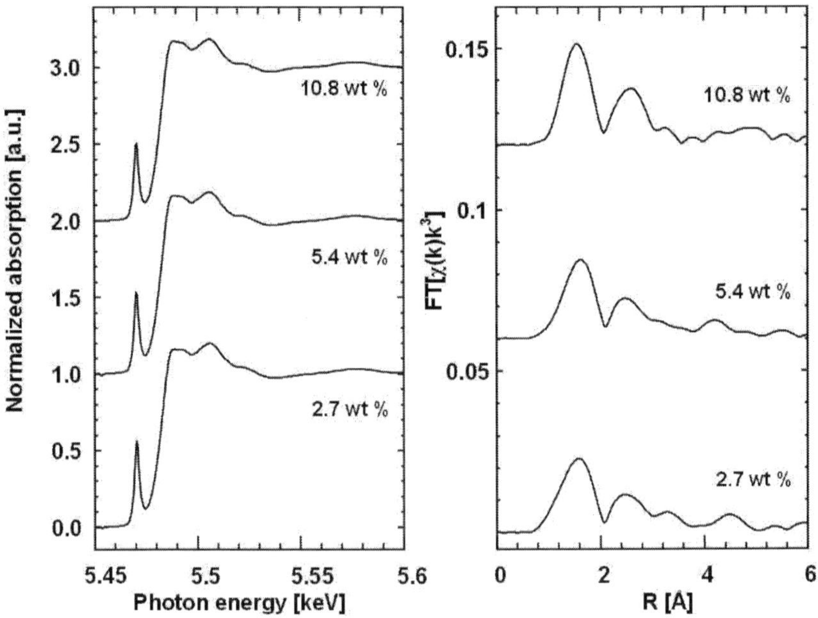

Figure 11. V K edge XANES spectra (a) and FT(χ(k)*k ^{3}) (b) of hydrated V_xO_y/SBA-15 samples with different vanadium loadings (2.7 wt %, 5.4 wt %, and 10.8 wt %).

As described above, the validity of the XAFS analysis approach chosen was evaluated by calculating confidence limits and F parameters (Table 5). The model structure employed corresponds to the local structure around V centers in bulk V_2O_5 (Table 6). In fitting procedure #1 there V-O distances (1.6 Å, 1.8 Å, and 2.0 Å) and two σ^2 (one for R = 1.6 Å and one for all other V-O distances) in the first V-O shell were allowed to vary independently. Additionally, three V-V distances (3.1 Å, 3.4 Å, and 3.6 Å) with the same σ^2 were refined. Moreover, E_0 was also allowed to vary in fitting procedure #1. Again because of N_{ind} = 18 and N_{free} = 10 refinement procedure #1 would be considered reliable according to the Nyquist criteria. Reasonable confidence limits and F = 0 were calculated for the V-V distances and σ^2(V-V) parameter. However, rather high confidence limits of the V-O distances of ± 0.05 Å and F parameters of 0.8 for both σ^2(V-O) parameters were obtained with procedure #1. Moreover, E_0 exhibited a confidence limit of ± 17.0 and F = 0.9. Hence, fitting procedure #1 clearly exceeds the number of meaningful parameters. Therefore, the fitting procedure was modified and the number of free parameters was reduced. E_0 was kept invariant again in the refinement, two V-O distances at ~1.6 Å and 1.9 Å, and one σ^2(V-O) parameter were used. In contrast to procedure #1, procedure #2 yielded reasonable confidence limits (e.g. ± 0.02 for V-O distances) and acceptable F parameters (mostly F = 0).

Table 5 indicates a small increase in the various V-O and V-V distances from V_2O_5 to hydrated V_xO_y/SBA-15. Intercalation of water in hydrated V_xO_y/SBA-15 may be accompanied by increasing nearest neighbor distances. More pronounced, though, is the increase in the disorder parameter σ^2 for the V-O and V-V scattering scattering paths used in the XAFS refinement for hydrated V_xO_y/SBA-15 (Table 5). In particular, V-V contributions are strongly damped in the FT(χ (k)*k^3) of hydrated V_xO_y/SBA-15 indicating an increased disorder in the local structure of hydrated V_xO_y species supported on SBA-15 compared to bulk V_2O_5.

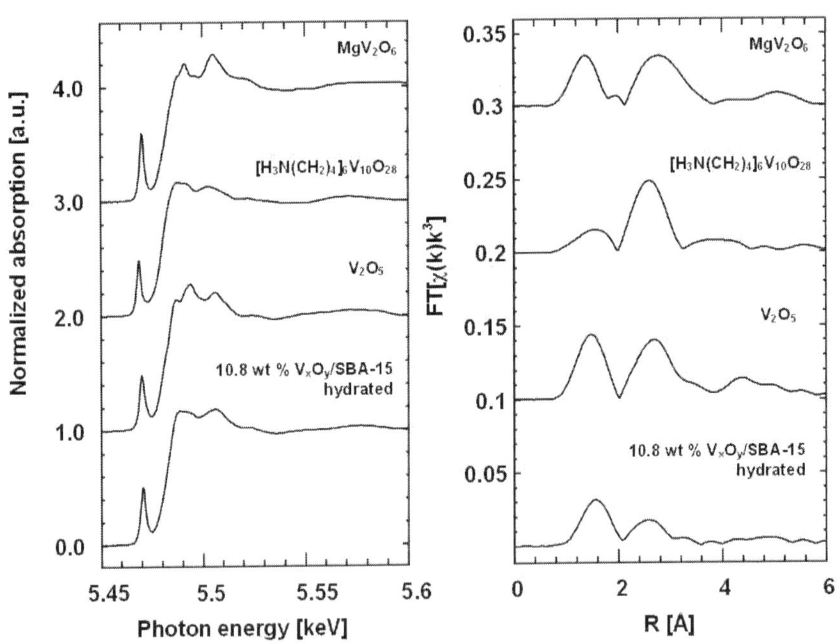

Figure 12. V K edge XANES spectra (a) and FT(χ(k)*k^3) of hydrated V_xO_y/SBA-15 (10.8 wt %) compared to those of various references (i.e. V_2O_5, $[H_3N(CH_2)_4]_6V_{10}O_{28}$, and MgV_2O_6).

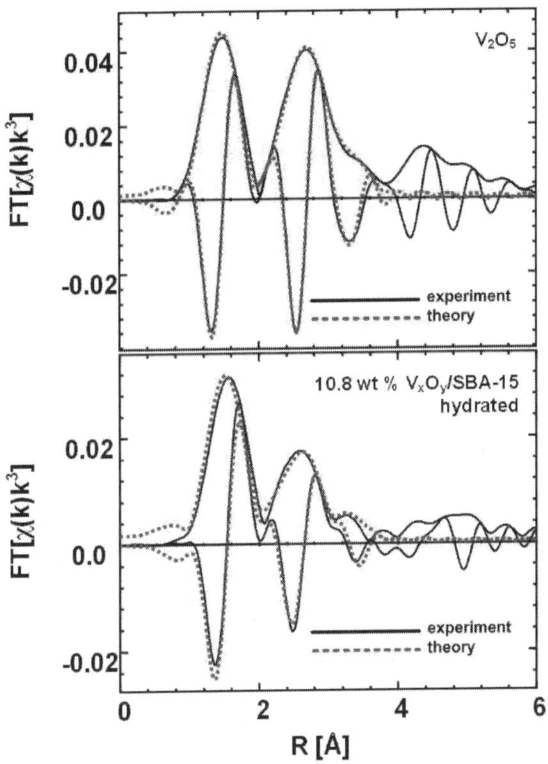

Figure 13. Experimental (solid) V K edge FT(χ(k)*k ^{3}) of hydrated V_xO_y /SBA-15 (10.8 wt %) (bottom) and of V_2O_5 (top) together with theoretical XAFS functions (V_2O_5 model).

The structural similarity between hydrated vanadium oxide species supported on SiO_2 and V_2O_5 has previously been observed by Raman spectroscopy [10]. Evidently, the local structure of hydrated V_xO_y/SBA-15 used here is very similar to other materials previously described in the literature. Dehydration should therefore result in a similar structure of the dehydrated phase. In addition to 10.8 wt % V_xO_y/SBA-15, samples with lower loadings of 2.7 wt % and 5.4 wt % V were measured (Figure 10and Figure 11) and analyzed according to the procedure described above. Very similar results were obtained for the hydrated state of the low-loading samples compared to 10.8 wt % hydrated V_xO_y/SBA-15. Apparently, in the range of V loadings from ~3 to 11 wt % the local structure of both hydrated and dehydrated V_xO_y/SBA-15 is largely independent of the amount of vanadium oxide supported on SBA-15.

Table 5 . Evaluation of EXAFS refinement of dehydrated V_xO_y/SBA-15.

Type	N	V_2O_5, procedure #1 Z	± z	F	V_2O_5, procedure #2 Z	± z	F	Hydrated V_xO_y-SBA-15 Z	± z	F
V - O	1	1.58	0.05	0.3	1.58	0.011	0	1.65	0.014	0
σ^2(V-O)	3(1)	0.011	0.0055	0.8	0.0104	0.0069	0	0.0126	0.0003	0
V - O	3(2)	1.87	0.06	0.3	1.88	0.014		1.92	0.018	0
σ^2(V-O)	-(3)	0.0098	0.004	0.8	-	-	-	-	-	-
V - O	-(1)	1.93	0.03	0.5	-	-	-	-	-	-
V - V	2	3.12	0.006	0	3.11	0.003	0	3.08	0.0075	0
σ^2(V-V)	3	0.0047	0.00032	0	0.0047	0.00035	0	0.013	0.0006	0
V - V	1	3.39	0.044	0	3.38	0.046	0	3.42	0.046	0.3
V - V	2	3.60	0.024	0	3.59	0.024	0	3.65	0.028	0
E_0	-	0.38	17.0	0.9	-	-	-	-	-	-

V K edge XAFS parameters (Z for distances R and disorder parameter σ^2) obtained from two different procedures of fitting a model structure (i.e. V_2O_5) to the experimental XAFS FT(χ(k)*k^3) of reference V_2O_5 and hydrated V_xO_y/SBA-15 (10.8 wt %) (details of fit given in Table 6) together with confidence limits (± z, referring to 95% of fit residual) and significance parameter F (details given in text). Fit residual 6.0 for V_2O_5 procedure #1, 6.3 for V_2O_5 procedure #2, and 13.6 for hydrated V_xO_y-SBA-15.

Table 6. EXAFS refinement results obtained for experimental FT(χ(k)*k^3) of hydrated V_xO_y/SBA-15 and V_2O_5.

		R_{Model} [Å]	V_2O_5		hydrated V_xO_y-SBA-15	
Type	N	R_{Model} [Å]	R [Å]	σ^2 [Å^2]	R [Å]	σ^2 [Å^2]
V - O	1	1.58	1.58	0.0104	1.65	0.0126
V - O	1	1.78	1.88	0.0104_C	1.92	0.0126_C
V - O	2	1.88	1.88_C	0.0104_C	1.92_C	0.0126_C
V - O	1	2.02	1.88_C	0.0104_C	1.92_C	0.0126_C
V - V	2	3.08	3.11	0.0047	3.08	0.0129
V - V	1	3.43	3.39	0.0047_C	3.43	0.0129_C
V - V	2	3.56	3.59	0.0047_C	3.65	0.0129_C

Type and number (N) of atoms at distance R from the V atoms in a V_2O_5 system compared to experimental distances and XAFS disorder parameter (σ^2). Parameters were obtained from refinement of a V_2O_5 model structure (ICSD 60767) to the experimental V K edge XAFS FT(χ(k)*k^3) of hydrated V_xO_y/SBA-15 (10.8 wt %) and bulk V_2O_5(). (k range from 2.7 - 11.0 Å^{-1}, R range 0.9 - 3.8 Å, E_0 (V_xO_y/SBA-15) = 0.0 eV/E_0 (V_2O_5) = 0.0, fit residual 13.4 (V_xO_y/SBA-15) and 6.3 (V_2O_5), N_{ind} = 18, N_{free} = 7) (Subscript C indicates parameters that were correlated in the refinement). Confidence limits and significance of fitting parameters are given in Table 5.

3. CONCLUSIONS

X-ray absorption spectroscopy is a very suitable technique for studying the local structure of dispersed metals or metal oxides on various support materials. Conventional XAFS analysis consists of finding a suitable model structure and fitting the corresponding theoretical XAFS functions to the experimental data. Because the number of potential parameters often exceeds the number of "independent" parameters, evaluating the reliability and significance of a particular fitting procedure is mandatory. Therefore, the number of independent parameters (Nyquist) alone is not sufficient. Here, we have employed confidence limits and F parameters to identify suitable analysis procedures. The local structure of vanadium oxide supported on nanostructured SiO_2 (SBA-15) was investigated. Three samples with different vanadium loadings (i.e. 2.7 wt %, 5.4 wt %, and 10.8 wt %) were employed. Thermal treatment in air at 623 K resulted in characteristic structural changes of the V oxide species. The local structure of dehydrated V_xO_y/SBA-15 was described best by assuming a model structure consisting of an ordered arrangement of neighboring V_2O_7 units. This is in good agreement with recent NEXAFS studies and theoretical calculations that also concluded the presence of V-V bonds in the V_xO_y species supported on SBA-15 [28]. Moreover, the local structures of both hydrated and dehydrated V_xO_y/SBA-15 were found to be independent of the V loading over the range employed. With respect to XAFS data previously presented in the literature and interpreted in terms of isolated VO_4 units it can be suggested that including contributions of higher shells would also lead to conclude polymeric V_xO_y units. Comparing the influence of surface properties and structure of various support materials will be the subject of future work. Eventually, onset of catalytic activity in selective propene oxidation was detected at about 573 K. In situ XAS data measured under reaction conditions indicated, that the characteristic ordered arrangement of V_2O_7 dimers in the local structure of dehydrated V_xO_y/SBA-15 persisted. Future studies on V_xO_y species supported on SBA-15 as model systems for vanadium based selective oxidation catalysts will have to take the presence of V_2O_7 species rather than isolated VO_4 units into account.

4. EXPERIMENTAL

4.1 Sample preparation

Silica SBA-15 was prepared according to literature procedures [2]. Details of preparation and characterization of the same catalysts were described elsewhere [9, 10, 14]. Briefly, SBA-15 was functionalized by adding 3-aminopropyltrimethoxysilane (APTMS) to a suspension of SBA-15 in toluene at 338 K. The suspension was stirred for 12 hours. The contents were filtered, washed and finally stirred in 0.3 M HCl for 12 hours. The contents was filtered again, washed with water and dried in air overnight (functionalized SBA-15). The vanadium oxides supported on SBA-15 were prepared by adding appropriate amounts of butylammonium decavanadate [31] to a suspension of functionalized

SBA-15 in water. The resulting powder was calcined at 823 K for 12 hours. The results of the N_2 physisorption analysis of the SBA-15 and the SBA-15 supported vanadium oxide samples are given in and have been discussed in detail previously together with detailed structural characterization [10]. Importantly, in the presence of vanadium oxide the hexagonal structure of SBA-15 is preserved, the mesopores remain accessible to reactants, and the vanadia species are located inside the pores of SBA-15.

The vanadium oxides supported on SBA-15 obtained are denoted as hydrated V_xO_y/SBA-15 (as-prepared) or dehydrated V_xO_y/SBA-15 (after thermal treatment). In addition to V_xO_y/SBA-15 several vanadium oxide reference compounds with an average valance of +5 were measured. These were either used as-purchased (V_2O_5 (Alfa Aesar 99.8%), NH_4VO_3 (Riedel de Haën, 99.5%) and Na_3VO_4 (Alfa Aesar, 99.9%)) or were prepared according to literature procedures ($Mg_2V_2O_7$, MgV_2O_6, $Mg_3V_2O_8$ [32], and $[H_3N(CH_2)_4]_6V_{10}O_{28}$ [31]).

4.2 X-ray absorption spectroscopy (XAS)

In situ transmission XAS experiments were performed at the V K edge (5.465 keV) at beamline E4 at the Hamburg Synchrotron Radiation Laboratory, HASYLAB, using a Si(111) double crystal monochromator. The energy range used for V K near edge scans (XANES) and extended XAFS scans (EXAFS) was 5.4-5.7 keV (~3 min/scan) and 5.4-6.0 keV (~20 min/scan), respectively. For in situ and ex situ XAFS measurements samples were mixed with BN and PE, respectively, and pressed into self-supporting pellets (5 mm and 13 mm in diameter, respectively). In order to obtain an edge jump, $\Delta \mu_x$, at the V K below 1.0, 2 mg of 10.8 wt % and 5.4 wt % V_xO_y/SBA-15, 3 mg of 2.7 wt % V_xO_y/SBA-15, ~1 mg for bulk vanadium oxides (NH_4VO_3, V_2O_5, $[H_3N(CH_2)_4]_6V_{10}O_{28}$, Na_3VO_4) diluted with BN (~15 mg), and 3-6 mg for bulk vanadium oxides ($Mg_3V_2O_8$, MgV_2O_6) diluted with PE (~200 mg) were employed. Transmission XAS measurements were performed in an in situ cell described previously [33]. Dehydration of V_xO_y/SBA-15 was conducted in 20% O_2 and He (total flow 30 ml/min) in a temperature range from 293 K to 623 K at a heating rate of 5 K/min and a holding time of 30 min at 623 K. Reaction tests were performed in 5% propene and 6% O_2 in He in the temperature range from 293 K to 723 K (5 K/min, total flow 30 ml/min). The gas atmosphere was analyzed using a noncalibrated mass spectrometer in a multiple ion detection mode (QMS200 from Pfeiffer). Ex situ XAFS measurements were performed in He atmosphere at room temperature.

X-ray absorption fine structure (XAFS) analysis was performed using the software package WinXAS v3.2 [34]. Background subtraction and normalization were carried out by fitting linear polynomials to the pre-edge and 3rd degree polynomials to the post-edge region of an absorption spectrum, respectively. The extended X-ray absorption fine structure (EXAFS) $\chi(k)$ was extracted by using cubic splines to obtain a smooth atomic background μ_0 (k). The FT($\chi(k)*k^3$), often referred to as pseudo radial distribution function, was calculated by Fourier transforming the k^3-weighted experimental $\chi(k)$ function, multiplied by a Bessel window, into the R space. EXAFS data analysis was performed using theoretical backscattering phases and amplitudes calculated with the ab-initio multiple-scattering code FEFF7 [35]. EXAFS refinements were performed in R space simultaneously to magnitude and imaginary part of a Fourier transformed k^3-weighted and k^1-weighted experimental χ (k) using the standard EXAFS formula [36]. This procedure strongly reduces the correlation between the various XAFS fitting parameters. Structural parameters allowed to vary in the refinement were (i) disorder parameter σ^2 of selected single-scattering paths assuming a symmetrical pair-distribution function and (ii) distances of selected single-scattering paths. Coordination numbers (CN), E_0 shifts, and amplitude reduction factor S_0^2 were kept invariant in the final fitting procedures. Correlations of specific parameters to reduce the number of free running parameters and to improve the stability of the refinement are described below.

The statistical significance of the fitting procedure employed was carefully evaluated in three steps. First, the number of independent parameters (N_{ind}) was calculated according to the Nyquist theorem $N_{ind} = 2/\pi*\Delta R* \Delta k + 2$. In all cases the number of free running parameters in the refinements was well below N_{ind}. Second, confidence limits were calculated for each individual parameter. In the corresponding procedure, one parameter was successively varied by a certain percentage (i.e. 0.05% for R and 5% for σ^2) and the refinement was restarted with this parameter kept invariant. The parameter was repeatedly increased or decreased until the fit residual exceed the original fit residual by more than 5%. Eventually, the confidence limit of the parameter was obtained from linear interpolation between the last and second last increment for an increase in fit residual of 5%. This procedure was consecutively performed for each fitting parameter. Third, a so-called F test was performed to assess the significance of the effect of additional fitting parameters on the fit residual. The corresponding procedure was adopted from the well-known library "Numerical Recipes in C" where it is described in detail [37]. In short, one parameter was varied by a certain percentage (i.e. between 2 and 8% for R and between 10 and 80% for σ^2) and the refinement was restarted with this parameter kept invariant. Subsequently, the difference between experimental and theoretical function (i.e. magnitude and imaginary part of $FT(\chi(k)*k^3)$ for a refinement in R space) was calculated and compared to that of the original refinement. The corresponding F parameter ranges between 0.0 and 1.0, where F = 1.0 indicates an insignificant change in the fit residual, while F = 0.0 indicates a highly significant change in fit residual. The iterative procedure was terminated when the corresponding F parameter was below 0.7. Fit parameters with F = 0.8 or higher are most likely strongly correlated and may be statistically insignificant. These parameters should be kept invariant in the refinement. Eventually, this procedure was also consecutively performed for each fitting parameter.

ACKNOWLEDGEMENTS

The Hamburg Synchrotron Radiation Laboratory, HASYLAB, is acknowledged for providing beamtime for this work. This research was supported by the Deutsche Forschungsgemeinschaft (DFG). The authors acknowledge support from the Cluster of Excellence *"Unifying Concepts in Catalysis"* (DFG). C.H. thanks the DFG for an Emmy Noether fellowship (DFG).

COMPETING INTERESTS

The authors declare that they have no competing interests.

AUTHORS' CONTRIBUTIONS

AW participated in XAFS data measurements and analysis, and drafted the manuscript. RH performed the sample preparation and N2 adsorption analysis. CH participated in design and coordination of the study, sample preparation, and N2 adsorption analysis. TR participated in XAFS data measurements and analysis, and design and coordination of the study. All authors read and approved the final manuscript.

REFERENCES

1. Ressler T, Walter A, Huang ZD, Bensch W: Structure and properties of a supported MoO_3-SBA-15 catalyst for selective oxidation of propene. J Catal. 2008, 254: 170-179.
2. Zhao DY, Huo QS, Feng JL, Chmelka BF, Stucky GD: Nonionic Triblock and Star Diblock Copolymer and Oligomeric Surfactant Syntheses of Highly Ordered, Hydrothermally Stable, Mesoporous Silica Structures. J Am Chem Soc. 1998, 120: 6024-6036.
3. Hess C: Characterization of the synthesis and reactivity behavior of nanostructured vanadia model catalysts using XPS and vibrational spectroscopy. Surf Sci. 2006, 600: 3695-3701.
4. Hess C, Drake IJ, Hoefelmeyer JD, Tilley TD, Bell AT: Partial Oxidation of Methanol Over Highly Dispersed Vanadia Supported on Silica SBA-15. Catal Lett. 2005, 105: 1-8.
5. Liu YM, Cao Y, Yi N, Feng WL, Dai WL, Yan SR, He HY, Fan KN: Vanadium oxide supported on mesoporous SBA-15 as highly selective catalysts in the oxidative dehydrogenation of propane. J Catal. 2004, 224: 417-428.

6. Liu W, Lai SY, Dai H, Wang S, Sun H, Au CT: Oxidative dehydrogenation of n-butane over mesoporous VO_x/SBA-15 catalysts. Catal Lett. 2007, 113: 147-154.

7. Keller DE, Visser T, Soulimani F, Koningsberger DC, Weckhuysen BM: Hydration effects on the molecular structure of silica-supported vanadiumoxide catalysts: A combined IR, Raman, UV-vis and EXAFS study. Vib Spectrosc. 2007, 43: 140-151.

8. Venkov TV, Hess C, Jentoft FC: Redox Properties of Vanadium Ions in SBA-15-Supported Vanadium Oxide: An FTIR Spectroscopic Study. Langmuir. 2007, 23: 1768-1777.

9. Hess C, Wild U, Schlögl R: The mechanism for the controlled synthesis of highly dispersed vanadia supported on silica SBA-15. Microporous and Mesoporous Materials. 2006, 95: 339-349.

10. Hess C, Tzolova-Müller G, Herbert R: The Influence of Water on the Dispersion of Vanadia Supported on Silica SBA-15: A Combined XPS and Raman Study. J Phys Chem C. 2007, 111: 9471-9479.

11. Hess C: Direct correlation of the dispersion and structure in vanadium oxide supported on silica SBA-15. J Catal. 2007, 248: 120-123.

12. Oyama ST, Went GT, Lewis KB, Bell AT, Somorjai GA: Oxygen Chemisorption and Laser Raman Spectroscopy of Unsupported and Silica-Supported Vanadium Oxide Catalysts. J Phys Chem B. 1989, 93: 6786-6790.

13. Gao X, Bare SR, Weckhuysen BM, Wachs IE: In Situ Spectroscopic Investigation of Molecular Structures of Highly Dispersed Vanadium Oxide on Silica under Various Conditions. J Phys Chem B. 1998, 102: 10842-10852.

14. Hess C, Hoefelmeyer JD, Tilley TD: Spectroscopic Characterization of Highly Dispersed Vanadia Supported on SBA-15. J Phys Chem B. 2004, 108: 9703-9709.

15. Schraml-Marth M, Wokaun A: Spectroscopic Investigation of the Structure of Silica-supported Vanadium Oxide Catalysts at Submonolayer Coverages. J Chem Soc Faraday Trans. 1991, 87 (16): 2635-2646.

16. Gao X, Wachs IE: Investigation of Surface Structures of Supported Vanadium Oxide Catalysts by UV-vis-NIR Diffuse Reflectance Spectroscopy. J Phys Chem B. 2000, 104: 1261-1268.

17. Bronkema JL, Bell AT: Mechanistic Studies of Methanol Oxidation to Formaldehyde on Isolated Vanadate Sites Supported on MCM-48. J Phys Chem C. 2007, 111: 420-430.

18. Olthof B, Khodakov A, Bell AT, Iglesia E: Effects of Support Composition and Pretreatment Conditions on the Structure of Vanadia Dispersed on SiO_2, Al_2O_3, TiO_2, ZrO_2, and HfO_2. J Phys Chem B. 2000, 104: 1516-1528.

19. Keller DE, Koningsberger DC, Weckhuysen BM: Molecular Structure of a Supported VO_4 Cluster and Its Interfacial Geometry as a Function of the SiO_2, Nb_2O_5, and ZrO_2 Support. J Phys Chem B. 2006, 110: 14313-14325.

20. Keller DE, Koningsberger DC, Weckhuysen BM: Elucidation of the molecular structure of hydrated vanadium oxide species by X-ray absorption spectroscopy: correlation between the V ··· V coordination number and distance and the point of zero charge of the support oxide. Phys Chem Chem Phys. 2006, 8: 4814-4824.

21. Keller DE, Airaksinen SMK, Krause AO, Koningsberger DC, Weckhuysen BM: Atomic XAFS as a Tool To Probe the Reactivity of Metal Oxide Catalysts: Quantifying Metal Oxide Support Effects. J Am Chem Soc. 2007, 129: 3189-3197.

22. Keller DE, Weckhuysen BM, Koningsberger DC: Application of AXAFS Spectroscopy to Transition-Metal Oxides: Influence of the Nearest and Next Nearest Neighbour Shells in Vanadium Oxides. Chem--Eur J. 2007, 13: 5845-5856.

23. Weckhuysen BM, Keller DE: Chemistry, spectroscopy and the role of supported vanadium oxides in heterogeneous catalysis. Catal Today. 2003, 78: 25-46.

24. Tanaka T, Yamashita H, Tsuchitani R, Funabiki T, Yoshida S: X-Ray Absorption (EXAFS/XANES) Study of Supported Vanadium Oxide Catalysts. J Chem Soc Faraday Trans. 1988, 84 (9): 2987-2999.

25. Keller DE, de Groot FMF, Koningsberger DC, Weckhuysen BM: ËO4 Upside Down: A New Molecular Structure for Supported VO_4 Catalysts. J Phys Chem B. 2005, 109: 10223-10233.

26. Hawthorne FC, Calvo C: The crystal chemistry of the M^+VO_3 (M^+ = Li, Na, K, NH_4, Tl, Rb, and Cs) pyroxenes. J Solid State Chem. 1977, 22 (2): 157-170.

27. Gopal R, Calvo C: Crystal Structure of Magnesium Divanadate, $Mg_2V_2O_7$. Acta Crystallogr, Sect B. 1974, 30 (10): B2491-2493.

28. Cavalleri M, Hermann K, Knop-Gericke A, Hävecker M, Herbert R, Hess C, Oestereich A, Döbler J, Schlögl R: Analysis of silica-supported vanadia by X-ray absorption spectroscopy: Combined theoretical and experimental studies. J Catal. 2009, 262: 215-223.

29. Wong J, Lytle FW, Messmer RP, Maylotte DH: K-edge absorption spectra of selected vanadium compounds. Phys Rev B. 1984, 30: 5596-5610.

30. Enjalbert R, Galy J: A Refinement of the Structure of V_2O_5. Acta Crystallogr, Sect C. 1986, C42: 1467-1469.

31. Roman P, Aranzabe A, Luque A, Gutierrez-Zorilla JM: Preparation and solid state characterization of some alkylammonium decavanadates. Crystal structure of the hexakis(n-hexylammonium) decavanadate dihydrate. Mater Res Bull. 1991, 26: 731-740.

32. Gao X, Ruiz P, Xin Q, Guo X, Delmon B: Preparation and characterization of three pure magnesium vanadate phases as catalysts for selective oxidation of propane to propene. Catal Lett. 1994, 23: 321-337.

33. Ressler T, Jentoft RE, Wienold J, Günter MM, Timpe O: In Situ XAS and XRD Studies on the Formation of Mo Suboxides during Reduction of MoO_3. J Phys Chem B. 2000, 104: 6360-6370.

34. Ressler T: WinXAS: a Program for X-ray Absorption Spectroscopy Data Analysis under MS-Windows. J Synch Rad. 1998, 5: 118-122.

35. Rehr JJ, Booth CH, Bridges F, Zabinsky SI: X-ray-absorption fine structure in embedded atoms. Phys Rev B. 1994, 49: 12347-12350.

36. Ressler T, Brock SL, Wong J, Suib SL: Multiple-Scattering EXAFS Analysis of Tetraalkylammonium Manganese Oxide Colloids. J Phys Chem B. 1999, 103: 6407-6420.

37. Teukolsky A, Vetterling WT, Flannery BP: Numerical Recipes: The Art of Scientific Computing. 2007, Cambridge: Cambridge University Press, Third

CHAPTER 6

Quantifying uncertainty in measurement of mercury in suspended particulate matter by cold vapor technique using atomic absorption spectrometry with hydride generator

Nahar Singh1*, Tarushee Ahuja2, Vijay Narain Ojha1, Daya Soni1, S Swarupa Tripathy1 and Ivo Leito3

1CSIR-National Physical Laboratory, Dr K.S. Krishnan Marg, New Delhi 110012,India.

2 Department of Applied Chemistry, Delhi College of Engineering, University of Delhi, Bawana Road, Delhi 110042, India.

3 Institute of Chemistry, University of Tartu, Ravila 14a, Tartu 50411, Estonia.

ABSTRACT

As a result of rapid industrialization several chemical forms of organic and inorganic mercury are constantly introduced to the environment and affect humans and animals directly. All forms of mercury have toxic effects; therefore accurate measurement of mercury is of prime importance especially in suspended particulate matter (SPM) collected through high volume sampler (HVS). In the quantification of mercury in SPM samples several steps are involved from sampling to final result.

The quality, reliability and confidence level of the analyzed data depends upon the measurement uncertainty of the whole process. Evaluation of measurement uncertainty of results is one of the requirements of the standard ISO/IEC 17025:2005 (European Standard EN IS/ISO/IEC 17025:2005, issue1:1-28, 2006). In the presented study the uncertainty estimation in mercury determination in suspended particulate matter (SPM) has been carried out using cold vapor Atomic Absorption Spectrometer-Hydride Generator (AAS-HG) technique followed by wet chemical digestion process. For the calculation of uncertainty, we have considered many general potential sources of uncertainty. After the analysis of data of seven diverse sites of Delhi, it has been concluded that the mercury concentration varies from 1.59 ± 0.37 to 14.5 ± 2.9 ng/m^3 with 95% confidence level ($k = 2$).

KEYWORDS

Mercury, Atomic absorption, spectrometer-hydride generator, Uncertainty, evaluation, High volume sampler

1. INTRODUCTION

Mercury is a highly toxic element found as naturally and as a contaminant in the environment. Natural sources of atmospheric mercury include volcanoes, geologic deposits, and volatilization from the ocean, rocks, sediments, water, and soils. The toxic effects of mercury depend on its chemical form and the route of exposure. Methyl mercury [CH_3Hg^+] is the most toxic form, absorbed readily and excreted slower than other forms of mercury. It affects the immune system, damages the nervous system, alters genetic enzyme systems and also affects the senses of touch, taste, and sight.

There are several instrumental techniques like Atomic Fluorescence Spectrometry (AFS), inductively coupled plasma atomic emission spectroscopy (ICPAES), nevertheless, the cold vapor AAS is the best technique sensitivity wise and to take care of interference of the other species during measurements. Several researchers (Singh and Sarkar 2003; Walcerz et al. 1993; Erdem and Henden 2004 and Barth et al. 1992animals and human beings has been studied by Moore and Ramamoorthy 1984 and indicated the natural and anthropogenic sources through which it reaches in different medium. Laser ionization mass spectrometric investigations have shown that stratosphere contains 1–5 ng m^{-3} of mercury and out of which 90% mercury is present as gas-phase (Murphy et al. 1998; Schroeder and Munthe 1998) as elemental mercury. The concentrations of gaseous mercury and other trace metals in aerosols were measured at Hel site, Gulf of Gdansk (Bełdowska et al. 2006) during summer season. Elemental mercury vapor released from broken thermometers, causes tremors, gingivitis, and excitability when inhaled over a long period of time.

In chemical metrology most of the important decisions are based on the quality and the reliability of analytical results. This reliability, quality and confidence can be achieved by correct estimation of uncertainty of the measurements following the ISO/EURACHEM guidelines. In this regard ISO has published, "Guide to the expression of uncertainty in measurement (GUM)" in 1995 in collaboration with BIPM, IEC, IFCC, IUPAC, and OIML, which recently was revised as VIM: JCGM 200:2008, Evaluation of measurement data-Guide to the expression of uncertainty in measurement. In last several years the uncertainty estimations for Hg in different matrixes (Kayal et al. 2009; & Kayal and Singh 2010 Synek et al. 2000; Kumar and Riyazuddin, 2007) have been carried out by various authors. However, determination of mercury in suspended particulate matter (SPM) and evaluating the uncertainty value of the result is a difficult task as there are many parameters involved from sampling to final value, which influence the result. To the best of our knowledge there is no study published, which describes the determinations of mercury in SPM samples with the total uncertainty budget. So in this study, an attempt has been made for the evaluation and expression of uncertainty in measurement for the determination of mercury in environmental samples collected through HVS by AAS-HG, following the ISO GUM guideline (JCGM 100:2008) as well as additional relevant guidance documents (ISO ISO/DTS 21748:2003 and European Standard EN ISO/IEC 17025:2005).

2. EXPERIMENTAL DETAILS

2.1 Instruments and apparatus

The suspended particulate matter (SPM) sampling was carried out using a High-Volume Sampler (HVS-410, Make: Envirotech). SPM was collected on pre-desiccated cellulose filter papers. Before sampling, the filter papers were dried for 3 hrs in an oven at 105°C to remove moisture, if any, and to get constant weight. After drying, the filter papers were kept in a desiccator at room temperature. For weighing of sample Mettler Toledo make balance model AX 204 capacity 220 g has been used. Atomic Absorption Spectrometer (AAS) of Analytik Jena make Vario-6 with a hydride generator accessory was used for the analysis of mercury. The mercury measurements were carried out at 253.7 nm wavelength at optimum conditions of acids and reducing agent. The pipette of 5 mL and volumetric flask of 50 mL capacity used were of Borosil glass works, India Limited. The pipettes and volumetric flasks were calibrated prior to analysis by the National Physical Laboratory (National Metrology Institute of India) following international standard procedure and protocol. The reference standards stock solution of 1.0 mg/litre (BND 601.02) of mercury was used (prepared by NPL-India) after diluting up to the working range for the calibration of AAS-HG instrument. A reagent blank was also prepared and the correction applied wherever required. Mean absorbance value of the three measurements of two replicates of each sample was taken into consideration for the calculation of the concentration of mercury. All the acid digestion and dilution work was carried out in a cleaned laminar flow bench equipped with the proper exhaustive system.

2.2 Reagents

Hydrochloric acid (35%) and Nitric acid (69%), of GR grade (Guaranteed Reagent), which were further purified by sub boiling point distillation in a quartz glass device. Hydrogen peroxide (50%) and Sulphuric acid (98%) all of E. Merck (India) make were used. Deionized water (18.2 MΩ resistivity) prepared from Millipore milli-Q element water purification system, USA was used throughout the process.

2.3 Sampling sites and description

New Delhi is located 160 km south of the Himalayas at latitude 28°24' to 28°53'N and longitude 76°20'E to 77°20'E with an altitude of about 216 meter above mean sea level. Delhi's climate is mainly influenced by its inland position and the prevalence of continental air during major part of the year. Delhi has three distinct seasons: summer, monsoon and winter. In the proposed study seven sites were selected namely Ashram (S-1; Heavy traffic site), Azadpur (S-2; Industrial site), Loni Road (S-3; densely populated + traffic site), Pitampura (S-4; Ambient site), Highway no. 56 (S-5; Traffic site), Naraina (S-6; Industrial site) and NPL (S-7; Mixed site).

2.4 Sample preparation procedure

Two replicates of 18×11.5 cm^2 size (known weight at deposited area) of all the seven sites were taken for the determination of mercury into separately cleaned poly-tetrafluoroethylene (PTFE) vessels used in the microwave digestion system. To these vessels 15 mL of sub boiled nitric acid, 5 mL of hydrogen peroxide and 5 mL of DI water were added and closed tightly with PTFE vessel lid. The PTFE vessels containing the samples were kept for overnight at room temperature. After keeping the PTFE vessel overnight, 4 to 5 drops of sulphuric acid was added and the PTFE vessel was kept on hot plate by covering with Teflon lid at 100°C for 15–20 minutes and then heated again at 150°C. Then the lid of the PTFE vessel was opened and it was heated again after addition of 10 ml sub boiled concentrated hydrochloric acid. To ensure the complete removal of nitric acid the sample was heated once again with 5 mL hydrochloric acid. Finally the solutions were boiled with 15 ml 1:1 de-ionized water and sub boiled concentrated hydrochloric acid mixture. After digestion the whole content of the vessel was transferred into a 100 mL quartz centrifuge tube and it was centrifuged at 5000 rotations per minute for 2–3 minutes to separate solid particles from the solutions. The supernatant liquid was transferred into a glass beaker. The remaining residue in the vessel was washed several times with hot water to ensure complete transfer of digested sample from the vessel. The final solutions were made 50 mL by addition of de-ionized water. In this solution mercury was determined by AAS-HG using the respective standards and principal resonance line after reducing mercury in the presence of sodium borohydride (3%), sodium hydroxide (1%) and hydrochloric acid (1%) by cold vapor techniques (ambient room temperature). Mean absorbance values of six readings of two replicates (three each) were taken into the consideration for calculation of the concentration of mercury. Procedural blank for cellulose filter paper without sample was also done to check the blank levels for mercury and correction was applied wherever required.

2.5 Determination of recovery

For determining the recovery of the wet digestion procedure we have taken three separate SPM deposit filter papers of the same matrix assuming uniform deposition of mercury on the filter paper. The filter paper was divided in two parts (18 cm × 11.5 cm^2 each). One part was treated as such, while in the second part 10 ml of 1000 μg/L mercury solution was added in a PTFE vessel along with the filter paper and the remaining procedure was the same as in the case of test samples. The dilutions of the spiked samples were carried out in the calibration range of AAS-HG before analysis. When keeping samples for long time there is a danger of loss of mercury in the low concentration range. In the present study we analyzed all samples within one week after digestion. From our experience we know that when we analyze mercury by AAS-HG in the low concentration range, we have to prepare fresh standards every week (calibration range: 10 μg/L to 50 in μg/L). After one week, there is a loss of 5% to 7% in the predetermined value at 27 ± 3°C (ambient conditions of our lab). The concentration range of mercury in the samples was 15 μg/L to 30 μg/L. The concentration of hydrochloric acid in the final solution should be at least 4-5%. In 4% or higher concentration of hydrochloric acid there is no loss due to adsorption over glass substrate, however mercury may be lost due to evaporation and microbial growth within the solution.

2.6 Determination of concentration of mercury in SPM sample and its uncertainty

2.6.1 Measurement model

In the determination of mercury in SPM samples the major sources of the uncertainty have been included in the calculation of combined uncertainty according to the EURACHEM/GUM guidelines. Following EURACHEM/GUM guidelines concentration of mercury $C(M_{Hg})$ in the SPM samples has been evaluated using AAS-HG technique by following equation.

$$C(M_{Hg}) = \frac{C_{Hg} \times V_{TEST} \times W_{TSPM}}{V_{HVS} \times R_{HVS} \times W_{ANA}} = ng/m^3 \tag{1}$$

Where; $C(M_{Hg})$ = Concentration of mercury in ng/m^3; C_{Hg} = Concentration of mercury analyzed from AAS-HG in µg/litre; V_{TEST} = Volume made of the test sample after wet digestion process in mL; W_{TSPM} = Total weight of SPM deposit on 18 cm × 23 cm area; V_{HVS} = volume of the air processed through high volume sampler (HVS); W_{ANA} = Weight of SPM deposit on 18 cm × 11.5 cm area taken for wet digestion process. R_{HVS} = Total recovery of the method containing the contribution of HVS in capturing mercury and possible losses of mercury during sample preparation.

The evaluated values for the above factors are given in Table 1. The important parameters that contribute towards the uncertainty are also shown in the Cause and effect diagram in Figure 1. In accordance with GUM, the combined uncertainty for the mathematical model, which is in a product or quotient form, is given by:

$$\left[u_c(y)/y\right]^2 = \sum_{i=1}^{N}\left[p_i u(x_i)/x_i\right]^2 \tag{2}$$

Table 1 Various evaluated components for the determination of mercury in SPM samples

Sample I.D.	SPM (μg/m^3)	Concentration of mercury obtained from AAS-HG after reducing blank (μg/litre) [C_{Hg}]	Volume made for test samples (mL) [V_{TEST}]	Total weight of SPM deposited on 18 cm × 23 cm filter paper (g) [W_{TSPM}]	Weight of SPM deposited on 18 cm × 11.5 cm filter paper area taken for analysis (g) [W_{ANA}]	Volume of air processed through HVS at STP, (m^3) [V_{HVS}]	Recovery of the analytical procedure (%) [R_{HVS}]	**Concentration of mercury in ng/m^3 C(M_{Hg})
S-1	395	58.22 (2.82)*	50	0.17744	0.08872	493.52	94.39	12.55 ± 2.44
S-2	203	23.41 (2.13)*	50	0.09124	0.04562	414.97		6.00 ± 1.23
S-3	2337	62.32 (4.13)*	50	1.05086	0.52543	457.16		14.50 ± 2.87
S-4	251	6.83 (1.12)*	50	0.11279	0.05639	456.66		1.59 ± 0.37
S-5	338	14.04 (0.92)*	50	0.15214	0.07607	471.24		3.17 ± 0.63
S-6	429	23.11 (1.62)*	50	0.19300	0.09650	417.58		5.89 ± 1.17
S-7	347	16.6 (1.33)*	50	0.62306	0.31153	391.63		4.51 ± 0.91

*Value in parentheses is the standard deviation of six determinations of two replicates (three each).
**Concentration of metal mercury has been calculated by using eq.-1 and values reported with expanded uncertainty in ng/m^3.

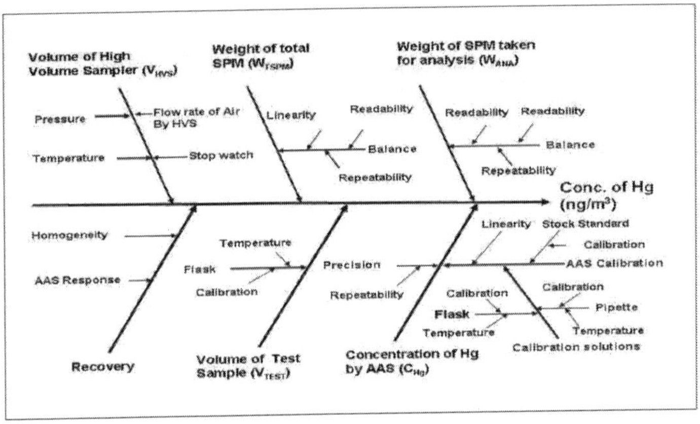

Fig. 1: Fish bone diagram for potential sources of uncertainty in mercury measurement by HG-AAS

Figure 1. Fish bone diagram for potential sources of uncertainty in mercury measurement by HG-AAS.

The combined uncertainty is given by:

$$\left[\frac{u_c(C(M_{Hg}))}{C(M_{Hg})}\right]^2 = \left[\frac{u(C_{Hg})}{C_{Hg}}\right]^2 + \left[\frac{u(V_{TEST})}{V_{TEST}}\right]^2 + \left[\frac{u(W_{TSPM})}{W_{TSPM}}\right]^2 + \left[\frac{u(W_{ANA})}{W_{ANA}}\right]^2 + \left[\frac{u(V_{HVS})}{V_{HVS}}\right]^2 + \left[\frac{u(R)}{R}\right]^2 \quad (3)$$

The uncertainty evaluation of all the input quantities is discussed in the following sub-sections.

2.6.2 Practicalities of uncertainty estimation

An uncertainty source may be 'Type A', which is evaluated by statistical analysis of a series of observations, or 'Type B', which is evaluated by using means other than the statistical analysis of a series of observations. To calculate standard uncertainty from the parameters of two most important distribution functions, if the limits $x \pm a$ are given without a confidence limit, it is appropriate to assume a rectangular distribution with a standard deviation of $a/\sqrt{3}$, but if values are given with confidence level, it is triangular distribution with a standard deviation of $a/\sqrt{6}$ (Ellison et al. 2000). Normal distribution is assumed when an estimate is made from repeated observation of a randomly varying process. In analytical chemistry, an expanded uncertainty (U) is used, which is obtained by multiplying the combined standard uncertainty, by a coverage factor k. The choice of coverage factor is based on the level of confidence. For an approximate level of confidence of 95%, the value of k is 2.

2.6.3 Uncertainty components in the measurement of concentration of mercury obtained from AAS-HG [C_{Hg}]

Referring to Figure 1, the major sources of the uncertainty for C_{Hg} are due to reference standard stock solution, repeatability, dilution in 50 mL volumetric flask from stock solution to working range and five mL capacity pipette used for dilution.

2.6.4 Reference standard stock solution

According to the certificate the concentration of the mercury standard stock solution is 1.00 ± 0.02 mg/L. Assuming normal distribution the standard uncertainty will be 0.01 mg/L. This stock solution was used for calibration of AAS-HG after proper dilution up to the working range.

2.6.5 Uncertainty evaluation due to repeatability

The repeatability in concentration of mercury was determined experimentally by analyzing six determinations of two independently digested replicates (three each). The standard deviation of six values (three each) has been used for calculation of standard uncertainty following EURACHEM guide.

2.6.6 Uncertainty in 50 ml volumetric flask used for dilution from 1000 ug/L to working range

The final volume of standard mercury solution made was 50 mL after several dilutions. The factors which include uncertainty in 50 ml volumetric flask are given in Figure 1. The major uncertainties were due to calibration and temperature. The uncertainty due to calibration and temperature has been calculated as below.

2.6.7 Uncertainty due to calibration

As per calibration certificate of supplier the expanded uncertainty for 50 mL flask is ± 0.02 mL at k = 2 assuming normal distribution, hence the standard uncertainty will be $u(V_{cal}) = 0.01$ mL.

2.6.8 Uncertainty due to temperature

According to calibration certificate the flask has been calibrated at 25°C and the laboratory temperature varies within the limits of ± 3°C. The volume expansion coefficient for water is 2.1×10^{-4}°C^{-1}, which leads to volume variation according to following equation.

$$\pm\left(V \times \Delta T \times 2.1 \times 10^{-4}\right) \qquad (4)$$

Where V is the volume of volumetric flask and ΔT is temperature variation in the laboratory. Volume variation for 50 mL flask $= \pm (50 \times 3 \times 2.1 \times 10^{-4}) = 0.0315\ mL$.

The standard uncertainty due to temperature effect, $u(V_T)$ is calculated using assumption of a rectangular distribution i.e. $u(V_T) = 0.0315/\sqrt{3} = 0.018$ mL.

The combined uncertainty for temperature and calibration in 50 mL volumetric flask has been calculated as;

$$u(VF) = \sqrt{[(0.01)^2 + (0.018)^2} = 0.021\ mL$$

2.6.9 Uncertainty due to five mL pipette used for dilution

5 mL aliquot was taken by 5 mL pipette for making dilution upto 50 mL. As per Figure 1 the main uncertainties were due to temperature variation and calibration accuracy (± 0.02 mL) assuming normal distribution as quoted by the manufacturer. The effect of temperature on 5 mL pipette has also been determined as per section, uncertainty in 50 ml volumetric flask used for dilution from 1000 ug/L to working range. The uncertainty of every dilution is different but this difference is negligible. Therefore for simplicity we have considered the uncertainty of first dilution even after performing the dilution second times.

$$u(pipette) = \sqrt{[(0.01)^2 + (0.002)^2} = 0.01\ mL$$

The Combined uncertainty due to measurement of concentration of mercury was calculated using data of subsection; uncertainty in reference standard stock solution, Uncertainty evaluation due to repeatability, Uncertainty in 50 ml volumetric flask used for dilution from 1000 ug/L to working range and Uncertainty due to five mL pipette used for dilution in following equation.

$$\left[\frac{u_c(C_{Hg})}{C_{Hg}}\right]^2 = \left[\frac{u(precision)}{X_{Precision}}\right]^2 + \left[\frac{u(cal\ std)}{X_{cal}}\right]^2 + \left[\frac{u(VF)}{V_{Flask}}\right]^2 + \left[\frac{u(pipette)}{V_{pipette}}\right]^2 \quad (5)$$

2.6.10 Uncertainty due to recovery

The recovery of mercury from the SPM filter paper depends on the storage of the samples and the method adopted for digestion. The loss of mercury in a closed system is less problematic during digestion process in comparison to an open system. The loss of mercury while keeping at low temperature i.e. $10 \pm 5°C$ has been found at our laboratory to be 15-20%, while at ambient temperature $30 \pm 5°C$ the loss of mercury was around 45–50% (both during one year period). For determining recovery of the wet digestion procedure we have taken three separate SPM deposit filter papers with the same matrix assuming uniform deposition of mercury on the filter paper. The mercury deposited filter paper was divided in two parts (18 cm × 11.5 cm each). One part was treated as such, while in the second part (200 µg/L) 10 ml of 1000 µg/L were added in the PTFE vessel along with filter paper and the remaining procedure was the same as in the case of test samples. The final solutions after digestion were made to 50 mL by de-ionized water. The dilutions of the spiked samples were carried out in the calibration range of AAS-HG before analysis. The recovery has been done as; Recovery % = Concentration of spiked sample/ (200 µg/L + concentration of non spiked sample). Following the same approach the average recovery of the spiked samples were found to be $94.39 \pm 0.80\%$.

2.6.11 Uncertainty evaluation in 50 mL volumetric flask used in making volume after wet chemical digestion [V_{TEST}]

The known area having known weight of the sample was processed through wet chemical route and the final volume was made to 50 mL.

Referring to Figure 1, the sources of uncertainty for V_{TEST} are due to calibration uncertainty and effect of temperature variation on 50 mL volumetric flask volume of aliquot taken. The uncertainty associated with temperature for 50 mL volumetric flask is carried out as per section 5.3. The combined uncertainty for V_{TEST} has been calculated on the basis of Figure 1 as follows;

$$u(V_{TEST})\sqrt{(0.01)^2 + (0.018)^2} = 0.021 \text{ ml}$$

2.6.12 Uncertainty evaluation due to volume of air processed through high volume sampler [V_{HVS}]

The volume of air sampled through HVS has been calculated on the basis of equation-6 and the volume of air processed at standard temperature and pressure (STP) is calculated by using equation-7 given below.

$$\begin{aligned}\text{Volume of air sampled} &= \textit{Time}\ (\text{min}) \\ &\quad \times \textit{Flow rate}\ (m^3/\text{min}) \\ &= 500 \times 1.1 = 550\, m^3\end{aligned} \qquad (6)$$

$$= \frac{V_1 \times P_1 \times T_1}{T_2 \times P_2} \qquad (7)$$

Where, V_1- Total volume of air processed through HVS, P_1- Atmospheric pressure on the date of measurement (mm of mercury); T_1- Standard temperature (273); T_2- working temperature [temperature of the sampling site °C + standard temperature] + standard temperature] and P_2- Standard pressure (760 mm of mercury).

Substituting value from Table 2 in eq.7

$$\text{Volume of air processed through HVS for S-1 at STP} \\ = \frac{550 \times 745.65 \times 273}{(25.5 + 273) \times 760} = 493.52\ (m^3)$$

Table 2. Uncertainty due to high volume sampler (**V** $_{HVS}$) for measuring volume of air at STP

Name of the site	Date of sampling	Sampling starting time	Sampling end time	Flow rate of air sampled through HVS (m^3/min)	Time (in min.)	Volume of air processed through HVS (m^3) V_1	Atmospheric pressure measured on site (mbar)	Atmospheric pressure in mm of mercury (mbar × 0.75) P_1	Average temperature of the sampling site (°C)	Average temperature of the sampling site (K) T_1	Volume of air processed through HVS at STP (m^3) [V_{HVS}]*
Heavy traffic site; Ashram zone (S-1)	31-1-08	9.10 am	5.30 pm	1.1	500	550	994.2	745.65	25.5	298.5	493.52
Industrial site; Azadpur zone (S-2);	1-2-08	10.0 am	5.00 pm	1.1	420	462	995.2	746.40	25.5	298.5	414.97
Densely populated + Traffic site Loni Road (S-3);	2-2-08	9.45 am	5.30 pm	1.1	465	511.5	993.6	745.20	26.5	299.5	457.16
Ambient site Pitampura zone (S-4);	3-2-08	9.45 am	5.30 pm	1.1	465	511.5	992.5	744.37	26.5	299.5	456.66
Traffic zone; Highway No. 56 (S-5);	4-2-08	10.30 am	6.30 pm	1.1	480	528	995.5	746.63	27.5	300.5	471.24
Industrial site; Nariana zone (S-6)	5-2-08	9.15 am	6.00 pm	1.1	425	467.5	996.3	747.22	27.5	300.5	417.58
Mixed site; NPL zone (S-7)	6-2-08	10.0 am	4.40 pm	1.1	400	440	992.8	744.60	27.5	300.5	391.63

*[V_{HVS}] is calculated using equation -7, following section uncertainty evaluation due to volume of air processed through high Volume sampler in section- 8.

Similarly V $_{HVS}$ for sample S-2 to S-7 has been calculated. Referring to Figure 1 the sources of the uncertainty for V $_{HVS}$ are flow rate of air through HVS, time recording by stop watch, ambient pressure measurement and measurement of temperature of the site. The uncertainty estimates of the parameters are taken from the calibration certificates. The sampling is carried out in open atmosphere so it is not possible to control the temperature of the atmosphere. So we have taken normal distribution for temperature variation, while the other analytical experiments were carried at controlled room temperature. The combined standard uncertainty for volume of HVS has been calculated using equation-8.

$$\left[\frac{u_c(V_{HVS})}{V_{HVS}}\right]^2 = \left[\frac{u(Flow)}{X_{Flow}}\right]^2 + \left[\frac{u(Time)}{X_{Time}}\right]^2 + \left[\frac{u(\text{Pressure})}{X_{\text{Pressure}}}\right]^2 + \left[\frac{u(Temp)}{V_{Temp}}\right]^2 \quad (8)$$

2.6.13 Uncertainty evaluation due to balance used for weighing the blank and SPM deposited filter paper [W_{TSPM}] and [W_{ANA}]

As per the certificate of balance the Linearity (mg) /readability (mg) / repeatability (mg); ± 0.2/0.01/0.07 has been used for calculation and the standard uncertainty has been calculated assuming rectangular distribution and dividing each by √3. The contribution needs to be counted twice, once for tare, and the other for gross weight. So the combined standard uncertainty due to balance is 0.00024 g.

The total weight of the SPM deposited on filter paper and weight of the filter paper used for wet digestion process was calculated as follows;

I. Total area of the cellulose filter paper for sampling =20.3 cm × 25.4 cm = 515.62 cm^2

II. Total area of the filter paper used for deposition (18 cm × 23 cm) = 414 cm^2

III. Total weight of the cellulose filter paper used for sampling ($515.6\ cm^2$) = 4.25504 g

IV. Total weight of the cellulose filter paper after deposition ($515.6\ cm^2$) = 4.47603 g

V. Total suspended particulate matter deposit on the filter paper ($515.6\ cm^2$) = (IV-III) = 0.22099 g

VI. Total suspended particulate matter deposit on the filter paper ($*414\ cm^2$); $W_{TSPM} =$

$$W_{TSPM} = \frac{0.22099 \times 414}{515.62} = 0.17744\,g$$

It has been considered that the thickness of the cellulose filter used is uniform throughout the area. On this basis we have calculated the weight of the deposit area, which is $414\ cm^2$.

VII. Weight of the SPM deposited area (18 cm × 11.5 cm) taken for analysis, $W_{ANA} = 0.08872$ g. The weight calculations for samples S-2 to S-7 were also done in same way as given above for sample-1.

2.6.14 Calculation of concentration and associated uncertainty of mercury for S-1; (ng/m^3)

The concentration of mercury in the seven SPM samples has been determined by AAS-HG using eq.1. Substituting the values from Table 1 into eq.1, the concentration of mercury (ng/m^3) in S-1 sample: $C(M_{Hg}) = \frac{58.22 \times 50 \times 0.17744 \times 100}{0.08872 \times 493.52 \times 94.39} = 12.55\,ng/m^3$.

The combined uncertainty in measurement of mercury have been calculated after putting values of various parameters from Table 3 (summary table) in eq.3.

$$\left[\frac{u_c(C(M_{Hg}))}{12.55}\right] = \sqrt{\left[\frac{1.30}{58.22}\right]^2 + \left[\frac{0.021}{50}\right]^2 + \left[\frac{0.00024}{0.17744}\right]^2 + \left[\frac{0.00024}{0.08872}\right]^2 + \left[\frac{46.54}{493.52}\right]^2 + \left[\frac{0.80}{94.39}\right]^2}$$

$$u_c\left[C(M_{Hg})\right] = 12.55 \times \sqrt{(0.00947)};\ u_c(M_{Hg})$$
$$= 12.55 \times 0.097 = 1.22;$$

$$
\begin{aligned}
U_{ex} &= 1.22 \times 2 \\
&= 2.44\,\text{ng/m}^3 (\text{at } 95\% \text{ confidence level; } k = 2)
\end{aligned}
$$

Table 3. Uncertainty components (summary table) for calculation of combined uncertainty

S.N.	$u(C_{Hg})/C_{Hg}$	$u(V_{TEST})/V_{TEST}$	$u(W_{TSPM})/W_{TSPM}$	$u(W_{ANA})/W_{ANA}$	$u(V_{HVS})/V_{HVS}$	$u(R)/R$
S-1	1.30/ 58.22	0.021/50	0.00024/ 0.17744	0.00024/ 0.08872	46.54/ 493.52	0.80/ 94.39
S-2	0.90/ 23.41	0.021/50	0.00024/ 0.09124	0.00024/ 0.04562	39.13/ 414.97	0.80/ 94.39
S-3	1.80/ 62.32	0.021/50	0.00024/ 1.05086	0.00024/ 0.52543	43.11/ 457.16	0.80/ 94.39
S-4	0.46/ 6.83	0.021/50	0.00024/ 0.11279	0.00024/ 0.05639	43.06/ 456.66	0.80/ 94.39
S-5	0.40/ 14.04	0.021/50	0.00024/ 0.15214	0.00024/ 0.07607	44.44/ 471.24	0.80/ 94.39
S-6	0.70/ 23.11	0.021/50	0.00024/ 0.19300	0.00024/ 0.09650	39.38/ 417.58	0.80/ 94.39
S-7	0.57/ 16.6	0.021/50	0.00024/ 0.62306	0.00024/ 0.31153	36.93/ 391.63	0.80/ 94.39

Therefore, Concentration of mercury in sample S-1 is = 12.5 ± 2.4 ng/m^3. Similarly concentration of mercury and combined uncertainty for mercury measurements have been calculated for samples S-2 to S-7. Table 4 shows the analytical results of the seven sites and a graphical representation of mercury concentration along with associated uncertainty is given in Figure 2.

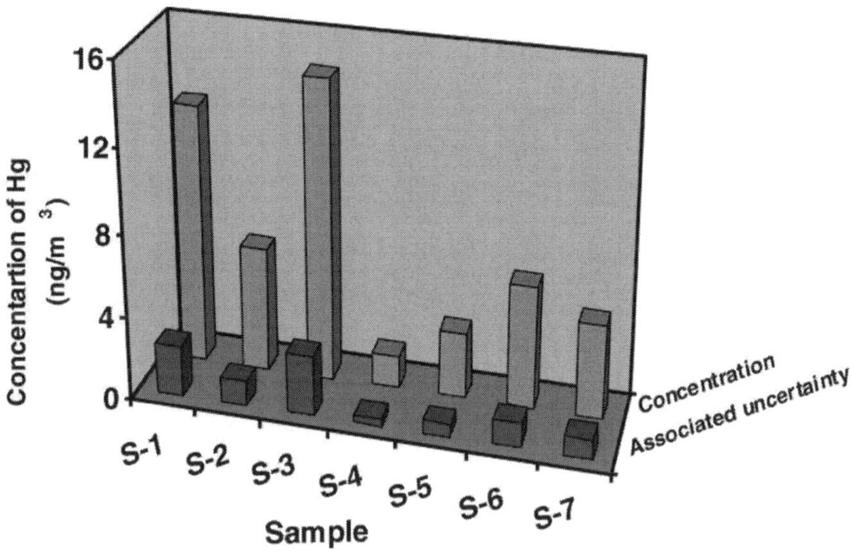

Figure 2. Concentration of each sample with associated uncertainty value.

3. RESULT AND DISCUSSION

Accurate determination of mercury in air is not easy due to the difficulties in measuring the air volume, losses during sample digestion process due to its vaporizing nature. The evaluation of combined uncertainty in mercury concentration in SPM sample is very complicated as there are various parameters, which contribute uncertainty in the entire process. On determining the uncertainty for mercury content measurement we were redundantly meticulous in identifying all uncertainty sources. In a routine determination of uncertainty we recommended that it is possible to be much less pedantic, and only reveal and realistically quantify the most significant components. The uncertainty components, which contribute a negligible value, could be ignored from the calculation of combined uncertainty value. The uncertainty components, which are found to contribute significantly during the subsequent computations, can be quantified more precisely and the final combined uncertainty can be recalculated. Mae and Dan 2010 also suggested how uncertainty can be reduced in measurement of atmospheric mercury and also suggested mercury in the air is measured as three forms like gaseous elemental mercury, reactive gaseous or gaseous oxidized and particle-bound mercury.

The concentration of mercury on the S-4 site is 1.57 ± 0.37 ng/m^3, which is the lowest (ambient site), while on the S-4 site (traffic/densely populated) the gighest concentration 14.5 ± 2.9 ng/m^3 was observed. The relative uncertainty in the measurement of mercury by FAAS-HG varies from 19 to 23%. Gajghate et al. 2011 have also estimated analytical uncertainty in quantification of mercury in different fractions of PM_{10} using microwave digestion and DMA-80 mercury analyzer. The mercury concentrations in ambient air in different various fractions of PM_{10} were found to be in the range of 3.13 ng m^{-3} (26%) – 22.96 ng m^{-3} (15%) with respective expanded relative uncertainty. Synek et al. 2000 estimated uncertainties in mercury determinations in biological materials by atomic absorption spectrometer – AMA 254 and suggested major uncertainty (52%) is due to sample absorbance measurement.

In the proposed determination of mercury concentration (S-1), there are five main uncertainty sources, which directly influence the final results those are C_{Hg}, V_{TEST}, W_{TSPM}, $W_{ANA.}$ V_{HVS} and R_{HVS}, with relative contributions 4.2%, 0.001%, 0.02%, 0.06%, 95.1% and 0.6% respectively. It is clear from the Figure 3 that flow rate (V_{HVS}) is the crucial parameter, which contributes the highest uncertainty of the value. The flow rate can be changed by several factors such as oily nature of the sample, presence of photochemical smog; wood smoke. These factors may block the filter paper and cause a rapid drop in air flow. Besides these high humidity and dense fog can cause the filter to become wet and thus airflow reduces through the filter paper. So sampling through HVS is the additional source, which also contributes uncertainty in mercury determination. Traffic and industry are the major sources, which contribute major contamination of mercury in environment.

Table 4 Results of mercury with uncertainty in SPM analyzed by AAS-HG

Name of the site	SPM in $\mu g/m^3$	Concentration of mercury in ng/m^3 with expanded uncertainty at 95% confidence; $k = 2$
Heavy traffic site (S-1); Ashram zone	395	12.6 ± 2.4
Industrial site (S-2) ; Azadpur zone	203	6.0 ± 1.2
Densely populated + Traffic site (S-3); Loni Road	2,337	14.5 ± 2.9
Ambient site (S-4) ; Pitampura zone	251	1.59 ± 0.37
Traffic site (S-5); Highway No.56	338	3.17 ± 0.63
Industrial site (S-6) ; Nariana zone	429	5.9 ± 1.2
Mix Zone (S-7); NPL zone	347	4.51 ± 0.91

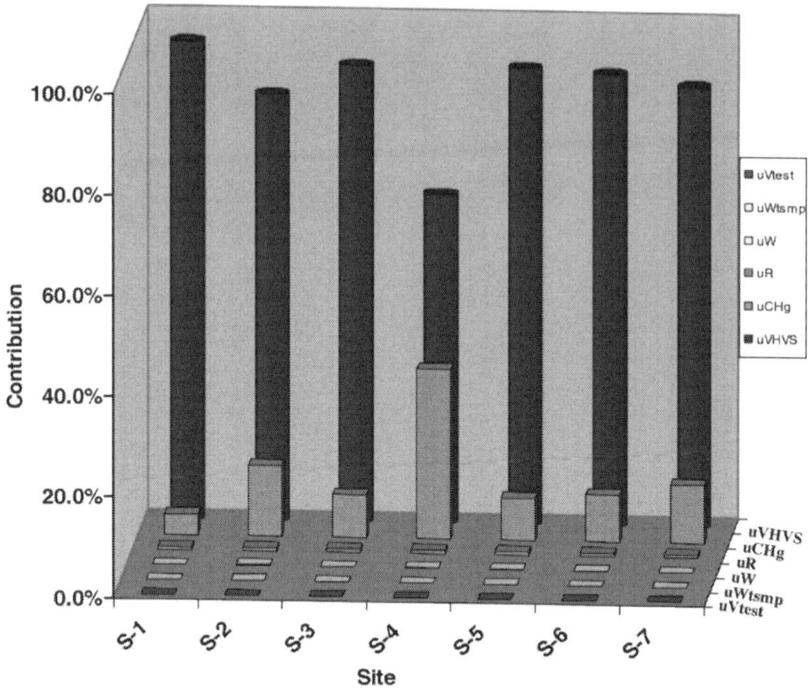

Figure 3. Contribution of each uncertainty source.

ACKNOWLEDGEMENT

The authors are grateful to Prof. R. C. Budhani; Director, National Physical Laboratory for encouragement and permission to publish this work. The authors are thankful to Sh Prabhat K. Gupta, Head Analytical Chemistry for his support. The authors also thank Dr. J. K. Bassin, Director, NEERI for providing their sites for sampling.

COMPETING INTERESTS

The authors declare that they have no competing interests.

AUTHORS' CONTRIBUTIONS

NS participated in the design of the study, carried out sampling, wet digestion of the sample and written work. TA and DS have contributed in analysis of the mercury analysis by AAS-HG. VNO and SST have contributed in calculation of various uncertainties, which have been identified in the experiments. IL has given valuable suggestion during writing of manuscript and also helped how to incorporate recovery of the process in the final calculation. All the authors have read and approved the final manuscript.

REFERENCES

1. Barth P, Krivan V, Hausbeck R: Cross-interferences of hydride-forming elements in hydride-generation atomic absorption spectrometry. Anal Chim Acta 1992, 263: 111-118.
2. Bełdowska M, Falkowska L, Lewandowska A: Airborne trace metals (Hg, Cd, Pb, Zn) of the coastal region, Gulf of Gdansk Oceanological and Hydrobiological Studies. Int J Oceanological Hydrobiology 2006, 5(2):159-169.
3. Ellison SLR, Rosslein M, Williams A: Quantifying Uncertainty in Analytical Measurement. EURACHEM /CITAC 2000, 1: 2000.
4. Erdem N, Henden E: Preconcentration and determination of inorganic arsenic using a multisyringe flow injection system and hydride generation-atomic fluorescence spectrometry. Anal Chim Acta 2004, 505: 59-65.
5. European Standard EN ISO/IEC 17025: General Requirements for the Competence of the Testing laboratories. Brussels: European Committee for Standardization; 2005. 14(2000) revised as ISO/IEC 17025
6. Gajghate DG, Pipalatkar PP, Khaparde VV, Rao CVC, Pustode TM: Uncertainty estimation in analysis of particulate-bound mercury in different size fractions of PM10 in ambient air. Accred Qual Assur 2011, 16: 459-465.

7. ISO ISO/DTS 21748: Guide to the Use of Repeatability, Reproducibility and Trueness Estimate. In Measurement Uncertainty Estimation. Geneva, Switzerland: ISO; 2003.

8. JCGM 100: Evaluation of measurement data-Guide to the expression of uncertainty in measurement. Geneva, Switzerland: JCGM guidance document is shared jointly by the JCGM member organizations (BIPM, IEC, IFCC, ILAC, ISO, IUPAC, IUPAP and OIML); 2008.

9. Kayal N, Singh N: Uncertainty calculation in the quantitative estimation of silica in rice husk ash by titrimetric method. Journal of MAPAN-Journal of Metrology Society of India 2010, 25(2):115-123.

10. Kayal N, Singh N, Ojha VN, Gupta PK: Evaluation and expression of uncertainty in the determination of alumina in deodorants by complexometric method. Test And Eva (ASTM) 2009, 37(4):321-328.

11. Kumar AR, Riyazuddin P: Underestimation of total arsenic concentration in groundwater samples determined by hydride generation quartz furnace atomic absorption spectrometry due to sample characteristics. Accred Qual Assur 2007, 12: 455-458.

12. Mae G, Dan J: Reducing the Uncertainty in Measurement and Understanding of Mercury in the Atmosphere. Environ Sci Technol 2010, 44: 2222-2227.

13. Moore W, Ramamoorthy S: Heavy metals in natural waters; Applied Monitoring and Impact Assessment. New York: Springer; 1984.

14. Murphy DM, Thomson DS, Mahoney MJ: In situ measurements of organics, meteoritic material, mercury, and other elements in aerosols at 5 to 19 kilometers. Science 1998, 282: 1664-1669.

15. Schroeder WH, Munthe J: Atmospheric Mercury, An overview. Atmos Environ 1998, 5: 809-822.

16. Singh N, Sarkar AK: Determination of Mercury in Fly ash ore by flameless cold vapor Atomic Absorption spectrometry using Hydride generator. Asian J Chem 2003, 159(3):1327-1330.

17. Synek V, Subrt P, Marecek J: Uncertainties of mercury determinations in biological materials using an atomic absorption spectrometer. Accred Qual Assur 2000, 5: 58-66.

18. VIM: JCGM 200: International vocabulary of metrology-basic and general concepts and associated terms (VIM). JCGM guidance document is shared jointly by the JCGM member organizations (BIPM, IEC, IFCC, ILAC, ISO, IUPAC, IUPAP and OIML); 2008.

19. Walcerz M, Bulska E, Hulanicki A: Study of some interfering processes in the arsenic, antimony and selenium determination by hydride generation atomic absorption spectrometry. Fresenius J Anal Chem 1993, 346: 622-626.

CHAPTER 7

Linde Type a Zeolite and Type Y Faujasite as a Solid-Phase for Lead, Cadmium, Nickel and Cobalt Preconcentration and Determination Using a Flow Injection System Coupled to Flame Atomic Absorption Spectrometry

Yaneira Petit de Peña[1], Wendy Rondón[2*]

[1]Laboratorio de Espectroscopía Molecular, Facultad de Ciencias, Universidad de Los Andes, Mérida, Venezuela

[2]Laboratorio de Química Analítica, Centro de Química, Instituto Venezolano de Investigaciones Científicas, Caracas, Venezuela

ABSTRACT

In this work, a flow injection analysis (FIA) method for the trace determination of lead, cadmium, nickel and cobalt in natural waters by formation of neutral chelates with ammonium pyrrolidine dithiocarbamate (APDC) was developed. The neutral chelates formed was retained in a mini-column packed with Linde type A zeolite (LTA) and type Y Faujasite zeolite (FAU) and then eluted with methyl isobutyl ketone (MIBK) to flame atomic absorption spectrometry (EAA) for its detection. Physicochemical characterization of this zeolite was carried out by Fourier Transform infrared spectroscopy and attenuated total reflectance (FTIR and IR-ATR), scanning electron microscopy and energy dispersive X-ray microanalysis (SEM-EDX) and X-ray power diffraction (XRD). Then, a FIA configuration was used with a column preconcentration system coupled to the detection system at room temperature (22°C). The detection limit and the relative standard deviation for 5 determinations of different solutions of Pb^{2+}, Co^{2+}, Ni^{2+} and Cd^{2+} for FAU and LTA zeolite were calculated. The sampling frequency ranged from 18 - 35 h^{-1} and preconcentration factors from 21 - 250 were achieved, for a sample volume of 6 mL using 20 mg of sorbents, indicating a high retention of the analytes on the zeolites material. The recoveries obtained in natural waters samples were close to 100% for all ions metal using synthetic zeolites, confirming the applicability of the method. The isotherm models of Langmuir, Scatchard, Freundlich and Dubinin-Radushkevich were used to study the equilibrium data, indicating that successfully followed the Freundlich and Dubinin-Radushkevich (D-R) isotherms at low metal ion concentration. The Freundlich parameter n varied between 0.35 - 1.01, whereas D-R isotherm yields the sorption free energy $E < 8$ $kJ \cdot mol^{-1}$ indicating psysisorption.

KEYWORDS

Cobalt; Nickel; Cadmium; Lead; Flow Injection; Preconcentration; Zeolites; Atomic Absorption

1. INTRODUCTION

The selective extraction and determination of heavy metals, especially some toxic ones, are still an intensively active research areas due to their biological mechanisms. The traditional methods for noble metals purification were precipitation and complexation procedure which needed large numbers of toxic acids and complexation agents, such as potassium or sodium cyanides that have negative impact on environment

[1]. Analytical techniques such as inductively coupled plasma optical emission spectrometry (ICP-OES) [2], and inductively coupled plasma mass spectrometry (ICP-MS) [3] are available for the determination of trace metals with sufficient sensitivity for most of applications, but the high cost of maintenance and correction of inherent interferences is usually a limitation [4]. On the other hand, flame atomic absorption spectrometry (FAAS) is the most widely used analytical method in this field due to its low cost, friendly operation, high sample throughput and good selectivity. However, there are some drawbacks that lessen sensitivity of the technique, including low sample introduction efficiency and low residence time of the atoms in flame [5].

Therefore, analyte preconcentration is required. In this manner a large number of techniques for the preconcentration of lead, cadmium, cobalt and nickel, including liquid-liquid extraction [6], coprecipitation [7], microextraction [8], solid phase extraction [9], etc., have been developed [10]. The combination of flow injection analysis with atomic spectrometry (FIA-FAAS) has considerably extended capabilities of conventional atomic spectrometric methods in terms of efficiency, sensitivity, economy and freedom from interferences [11,12]. Also FIA configuration may contain mini-columns with sorbent material for the analytes preconcentration and improve selectivity in the analysis. For that reason new packing materials are still being proposed and examined [11]. Amberlites (XAD) [13,14], silica gel [15,16], bio-adsorbents [17,18], carbon nanotubes [19,20] and zeolites [21,22] have been used for preconcentration of trace metals [23], after complex formation of metal to facilitate retention on the adsorbent material., such as 8-hidroquinoline [24], APDC [21], 1-(2-Pyridylazo)- 2-naphthol (PAN) [25], ammonium diethyldithiophosphate (DDTP) [26], for ion exchange and/or complexation of metal ions has been applied to extraction-complexation of metal ions. Petit et al. [27] developed an on-line FI system for the preconcentration of Cu^{2+} onto a mini-column packed with synthetic zeolites; APDC was used to form neutral chelate of the metal, which is then eluted with MIBK and determined by FAAS. Preconcentration factors of copper ranging from 35 - 125 for Na-LTA and 30 - 65 for Na-FAU were readily achieved, related to the direct introduction of aqueous solutions into an atomic absorption spectrometer.

Zeolite [28] is an aluminum silicate that occurs both as natural and as produced synthetic. Zeolite has a threedimensional structure with pores. It consists of silicon, aluminum and oxygen ions. The silicon ions are neutrally charged in the crystal structure. Aluminum ions create negative places. To keep the charge in balance, a counterion (Na+, K+) or a proton (H+) is present in the pores. Some types of zeolite have just a large pore through the entire crystal structure, which is stipulated by the ring size. Changing the ratio of aluminum on silicon can also influence the pore size as well as the type of a counterion. All natural zeolites contain aluminum and are hydrophilic in nature [29,30]. The zeolites have great capacity for ionic exchange due to the charge equilibrium that attracts the closest cation, maintaining neutrality. The cationic exchange property is a function of the ratio of Si to Al. This capacity is expressed as the number of cations per mass or volume unit available for exchange [31]. Zeolites are mainly used as ionic exchangers (water softeners), molecular sieves, absorbents and catalysts. Many studies and investigations are carried out with different zeolites, due to its properties and characteristics [31]. The purpose of this work is to investigate the feasibility of adsorption of the neutral complex with the use of adsorption isotherms, as a modeling tool to describe the phenomenon that regulates the preconcentration (with separation from the matrix) of metal ions on sodium and calcium Y Faujasites (Na-FAU, Ca-FAU) and sodium and calcium A zeolite (Na-LTA and Ca-LTA) for designing a new sorbent material. Strong emphasis is devoted to study and evaluate the utility of the new sorbent to act as a selective solid phase extractor for separation and preconcentration of ultratrace amount of Pb^{2+}, Cd^{2+}, Ni^{2+} and Co^{2+}, in spiked natural water samples prior to the determination by FIA-FAAS.

2. EXPERIMENTAL

2.1. Apparatus

A flame atomic absorption spectrometer Varian model SpectrAA 110 (Sao Paolo, Brazil), equipped with a bead impact system in the burner chamber and deuterium as background correction was used throughout.

The flame composition was acetylene (flow rate: 2.0 $L \cdot min^{-1}$) and air (flow rate: 10.0 $L \cdot min^{-1}$). Nebulizer flow rate was 4.0 $mL \cdot min^{-1}$. Signals were obtained using SpectrAA-110 software, as peak heights. Absorbance measurements and currents were carried out at 217.0 nm and 4 mA for Pb, 232.0 nm and 4 mA for Ni, 228.8 nm and 4 mA for Cd and 240.7 nm and 7 mA for Co, using hollow cathode lamps Varian. The instrumental parameters were used according to the manufacturers recommendations. A pH meter Metrohm 701A (Ohio, USA) was used for pH adjustment.

A FIA system (Figure 1) equipped with: 1) one Gilson Minipuls-3 peristaltic pump (PP, Ohio, USA) four channels as liquid propulsion devices, equipped with tubes of polyvinyl chloride; two channels for sample and APDC reagent, one for water and one channel for the propulsion of MIBK, using the displacement bottle; 2) two reactions coils, assembled with politetrafluoroethylene (PTFE) tubing; 3) the sample injection was achieved using two six-way rotatory valve Rheodyne (Berlin, Germany) provided with fixed volume loops substituted by a reaction coil (MR) followed by a minicolum in valve V_2 and by another coil filled with MIBK in valve V_1. PTFE tubing (0.5 mm i.d.) was used for all connections. A GraLab900 (Ohio, USA) timer was used to select preconcentration/ elution steps.

2.2. Reagents and Samples

All chemicals reagent used were of analytical grade and ultrapure water from a Milli-Q (Barnstead NANO pureInfinity) to prepare all solutions. A 1000 $mg \cdot L^{-1}$ lead, nickel, cadmium and cobalt stock solutions were prepared by dissolving 1.000 g of the metal (Merck, Germany, 99.9% w/w), in a small volume of concentrated nitric acid (Fluka, Germany, 65.0% v/v, d = 1.42 $g \cdot L^{-1}$) and diluting to 1 L with 1% v/v HNO_3. Working solutions were prepared by dilution of the stock standard solution and were adjusted to optimum pH value with concentrated HNO_3. The aqueous solution of APDC (Sigma Chemical, Germany 99.0% w/w), using solid purchased was prepared daily in water. MIBK (Riedel de Haën, Germany, 99% v/v) was also used as eluent organic solvent. All bottles used for storing samples and standard solutions, as well as the glassware were washed in 10% v/v nitric acid for 24 h and finally rinsed with ultrapure water. The natural water samples were collected from Mérida City (Venezuelan) namely: tap water, ozonated water, Mucuy River and Albarregas River. All samples were filtered through 0.45 μm membrane filters, acidified to 0.03 mol·L^{-1} HNO_3 and stored at 4°C in acid-cleaned polyethylene bottles in order to determine the "dissolved metal" fraction.

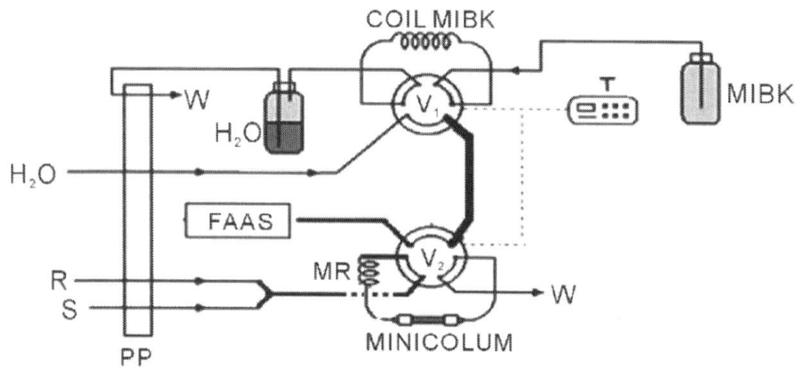

(A)

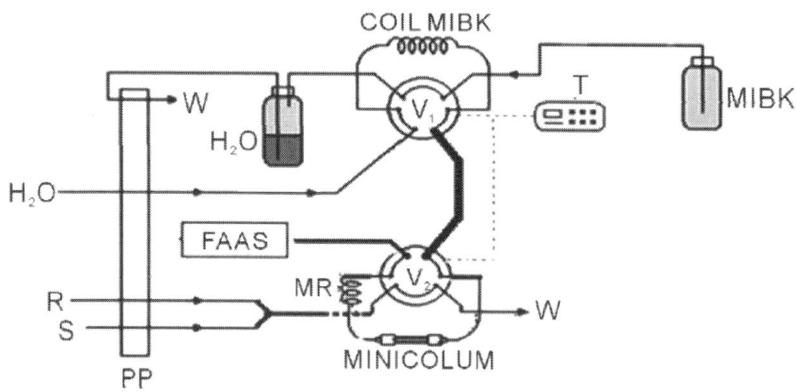

(B)

Figure 1. Flow injection manifold for preconcentration and determination of trace metal ions. R: APDC reagent.

2.3. Mini-Columns: Zeolite Synthesis, Construction and Packing

Zeolites synthetic (Na-LTA, Ca-LTA, Na-FAU, Ca-FAU) with 60 - 100 μm particle size (Millipore, Madrid, Spain) were employed as sorbent materials. The Na-FAU zeolite used in this work was synthesized in the Kinetic and Catalysis Laboratory of the Chemistry Department, Los Andes University. The raw material for the synthesis of Na-LTA zeolite (4A, [31]) used in this study was Venezuelan Kaolin from deposits located in Bolivar state (particle size less than 200 - 270 mesh) [32].

Kaolin was subject to metakaolinitation to produce a metakaolin. The metakaolin was then zeolitized, producing Na-LTA zeolite, following a procedure reported else where [31,32]. The Ca-LTA and Ca-FAU were obtained by ion exchange using 1.0 mol L^{-1} calcium chloride (Merck, Germany, 99.9% w/w) in a reflux system under temperature control. Amounts of 20 mg of the zeolites were used to pack the mini-columns. These were manufactured in the laboratory using PTFE tubing (4 mm id. and 0.3 cm length), sealed with a small piece of glass wool at the ends to avoid material losses and accommodated onto the FIA system. This configuration facilitated rapid replacement of the column, whenever required, thereby overcoming the deterioration of the analytical performance of the method due to the progressive deactivation of the sorbent material. The zeolites mini-column was proven to operate reliably for at least 200 sorption/elution cycles after washing with ultrapure water.

2.4. Characterization

The following tests were performed to characterize the synthesis and exchanges of Ca-FAU zeolite and Ca-LTA zeolite: FTIR, IR-ATR, SEM-EDX and XRD. FTIR and IR-ATR were performed using a FTIR Perkin Elmer (Germany) Spectrom RX1 and IR-ATR Perkin Elmer (Germany) model Spectrum 400, software-controlled Spectrum v.6.3.4.

The zeolites samples for testing were prepared using KBr pellet technique for FTIR and by the IR-ATR technique; the samples were placed directly into the diamond crystal with KRS-5 (mixture of bromide and thallium iodide) at ATR module. For the micrographs, it was used a scanning electron microscope (SEM) Hitachi S-2500 (Tokyo, Japan)coupled to an energy dispersive X-ray microanalyser Thermo Noran for the elemental analysis. XRD was performed using a Philips PW-1250 powder diffractometer equipped with an X-ray tube (Cu-Kα radiation, 40 kV, 25 mA). A small quantity of the sample was ground mechanically in an agate mortar, pestle and mounted on a flat holder covered with a thin layer of grease. Data collection was carried out in the 2θ range 5° - 60°, in steps of 0.02° and counting time of 10 s. Phase identification was performed by searching the ICDD powder diffraction file database, with the help of JCPDS (Joint Committee on Powder Diffraction Standards) files for inorganic compounds [33]. The relative intensity yields were obtained from normalized XRD intensities of the major reflection for each material.

2.5. The On-Line Mini-Column Preconcentration/FIA-FAAS

The manifold was operated in a mode based on time control (Figure 1) and the operational sequence is completed in two steps: preconcentration (load) and elution. In the first step (Figure 1(A)), 6 ml of the sample or standard solution containing 5 - 100 $\mu g{\cdot}L^{-1}$ $^{M2+}$ (at optimum pH for each metal M) were introduced into the system through the sample channel S, while 0.3% w/w APDC solution was through the reagent channel, R. These flows were continuously propelled using a peristaltic pump PP into the system through valve V_2 and mixed in the mixing reactor MR with the APDC, during optimum preconcentration time of each metal. Then the neutral chelate was adsorbed on the zeolite mini-column, located in the loop of valve V_2, and the sample matrix sent to waste. During this period, a stream of water was continuously passed through the nebulizer, so that the matrix of the sample never reached the detector. Meanwhile, the coil, located in the injection valve V_1 is charged with MIBK.

After the loading time, the injection valves were switched automatically by action of the timer to the injection step (Figure 1(B)) and the neutral chelate was eluted from the mini-column with MIBK directly into nebulizer and subsequently the flame. The aspiration flow rate of the nebulizer was 4.0 $\mu L \cdot min^{-1}$.

The peaks heights were used as analytical signals. The injection valves were again switched to initial position in order to have the system ready for the next preconcentration step.

2.6. Isotherms Adsorption Models

The adsorption experiments were performed on line, using the configuration without mini-colum to constant temperature of 22°C (Figure 2). The calibration curves in organic phase for each of the analytes under study were obtained following the next procedure: 1) Standards were prepared of 1 - 10 $mg \cdot L^{-1}$ Pb^{2+}, 0 - 3 $mg \cdot L^{-1}$ Cd^{2+}, 0 - 7 $mg \cdot L^{-1}$ de Ni^{2+} and 0 - 9 $mg \cdot L^{-1}$ Co^{2+} in MIBK containing 0.3% w/w APDC. The preparation in an organic solution, involves using small quantities in µL of aqueous solutions with high known concentrations of the analytes in study. 2) These organic solutions were loaded in the coil by the displacement system and then were turning the injection valve (Figure 2) to insert a stream of water that drove the analytes to the detector. Subsequently, establishing a mathematical relationship where the absorbances obtained by FIA system with columns of Figure 1 ($\mu g \cdot L^{-1}$ aqueous solutions, C_e) are substituted into the linear regression equation obtained with FIA system without column (organic solutions) of the Figure 2, for obtained the equivalent values in $ng \cdot g^{-1}$ of zeolite material, called q.

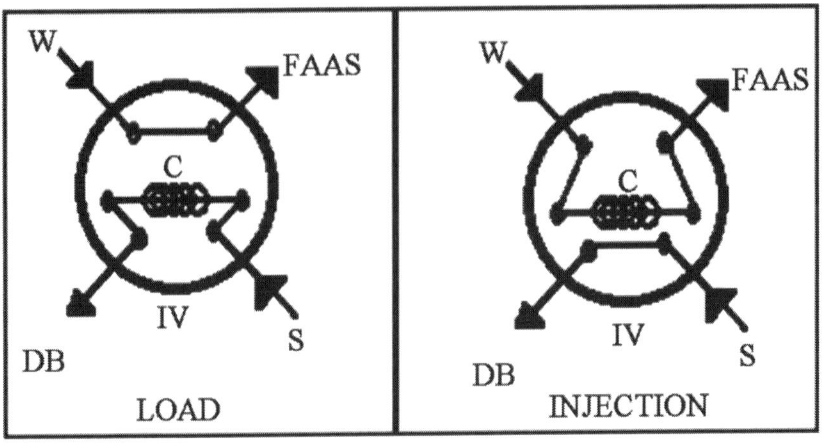

Figure 2. Injection valve setup (IV) controlled by timer. The flow propelled by peristaltic pump. S: metal ion solution in MIBK/APDC. C: organic solution coil. W: waste. DB: displacement bottle.

3. RESULTS AND DISCUSSIONS

3.1. Adsorbents Material Analysis

3.1.1. Spectra FTIR and ATR

Figure 3 shows the FTIR-spectra obtained for kaolin. The characteristic peaks at 3478 cm^{-1} and 3495 cm^{-1} are corresponding to the OH-stretching vibration. Bands at 1087 cm^{-1} and 1064 cm^{-1} were assigned to Si-O bonds in the SiO_4 molecules and Al-OH vibrations. The bands at 787 cm^{-1} and 641 cm^{-1} were Si-O symmetric stretching. Absorption at 480 cm^{-1} was assigned as Si-O-Al stretching vibration where the Al is in octahedral coordination [34]. On the other hand, comparing the FTIR-spectra of the obtained kaolin with the FTIR-spectra of Na-LTA and Na-FAU (Figure 4) one can appreciate differences between the bands of stretching vibrations and deformation of the H-O-H in the region of 3000 - 3700 cm^{-1} for the adsorbed molecular water which is located in the cavities of the LTA and FAU zeolites; and the band in 1650 cm^{-1} characterizes the signal stretching and angular deformation of the hydroxyl molecules of water of hydration of the zeolites. The bands in

1100 - 1050 cm^{-1} indicates the presence of Si-O and the crystallization bands appear in the zone of absorption 555 - 501 cm^{-1} which is assigned to the presence of structural double rings of LTA (D4R) and FAU (D6R) zeolites. As seen in Figure 5 corresponding to the ATR-spectra obtained for zeolites Na-LTA and Ca-LTA, appear weak bands in the zone 400 - 500 cm^{-1}, what are caused of vibration deformation of the tetrahedral layer being the area of 480 - 540 cm^{-1} particularly sensitive, where the frequency varies inversely with the ionic radius of octahedral cations.

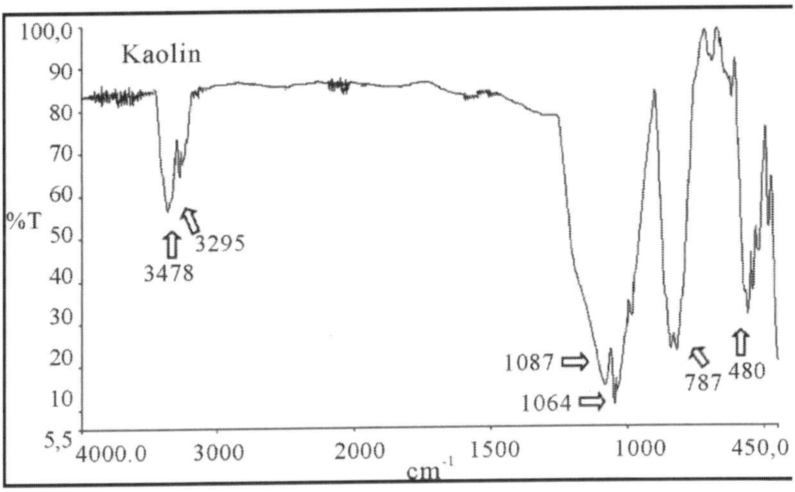

Figure 3. FTIR spectrum of kaolin.

3.1.2. SEM-EDX Analysis

In the SEM micrographs of Figure 6 it can be observed which of the samples of Na-LTA zeolites and Na-FAU zeolites presented greater amount of well formed cubic crystallites and hexagonal; typical morphology in these materials, and also shows that the crystals have dimensions of around 5 micrometers for Na-LTA zeolite and 3 micrometers for Na-FAU zeolite. The half elemental composition of the zeolite and Na-FAU were performed with three different random points.

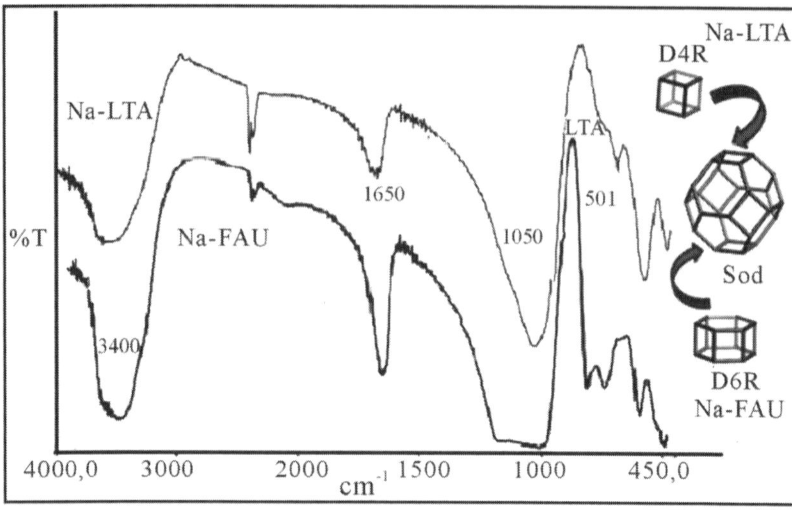

Figure 4. FTIR overlapping spectra of the synthesized zeolites: Na-LTA and Na-FAU.

The result obtained in EDX analysis with respect to Si/Al ranged around 1.3 for Na-LTA and 2.8 for Na-FAU which is an approximation of the corresponding value of the Si/Al obtained from the chemicals formulas $Na_{96}Al_{96}Si_{96}O_{384} \times 27H_2O$ [35] that has a Si/Al ratio equal to 1 for Na-LTA and $(M_{59/n})[(AlO_2)_{59}(SiO_2)_{133}] \times 235\ H_2O$ that has Si/Al ratio equal to 1.5 - 3.0 for Na-FAU [36].

3.1.3. XRD Aznalysis

XRD patterns of the Na-LTA and Na-FAU zeolites are reported in Figure 7" target="_self"> Figure 7. The powder XRD profiles are characteristic of highly crystalline material showing sharp reflections (2q positions).

The pattern of LTA (Figure 7(A)) shows all the characteristic peaks matching with the reported in the diffraction equipment library, with corresponding formula to $Na_2Al_2Si_{1,85}O_{7,7}\cdot 5,1\ H_2O$ (Ref. Code 00-038-0241) [33]. Also, is identified a small amount of quartz SiO_2 (Ref. Code

01-079-1906) [33]. The pattern of FAU (Figure 7(B)) shows all the characteristic peaks matching with the reported in the diffraction equipment library, with corresponding formula to $Na_2Al_2Si_{4,5}O_{13} \cdot H_2O$ (Ref. Code 00-043-0168) [33].

3.2. Optimization of Experimental Parameters

The flow system parameters were optimized by varying chemical and flow conditions by univariate manner. The analytical steps of the optimized procedure (FIA-FAAS) for lead, cadmium, nickel and cobalt determination are summarized in Table 1" target="_self"> Table 1.

3.3. Analytical Features and Analysis of Drinking Water Samples

Under the optimum conditions described above, the analytical parameters of the proposed method are summarized in Table 2 for each analyte. The experimental preconcentration factor was calculated as the radio of the slopes of calibration curves (SCC), obtained with and without preconcentration for the aqueous solutions (SCC of FIA-FAAS × 10^3/SCC of FAAS). The precision of the method was determined by seven repetitive analyses of different standard solution under the optimum conditions.

A good relative standard deviation in the range of 0.2% - 1.6% at 25 $\mu g \cdot L^{-1}$ Pb^{+2}, 40 $\mu g \cdot L^{-1}$ Cd^{+2}, 50 $\mu g \cdot L^{-1}$ Ni^{+2} and 40 $\mu g \cdot L^{-1}$ Co^{+2} (n = 5) was obtained. Recovery tests using the proposed method were performed using five different samples, and the test for each sample was carried out in triplicate. The preconcentration procedure was evaluated by spiking certain amounts of trace metals in natural water samples. The results for the recoveries of spiked samples are good since the values ranged from 98% - 102%.

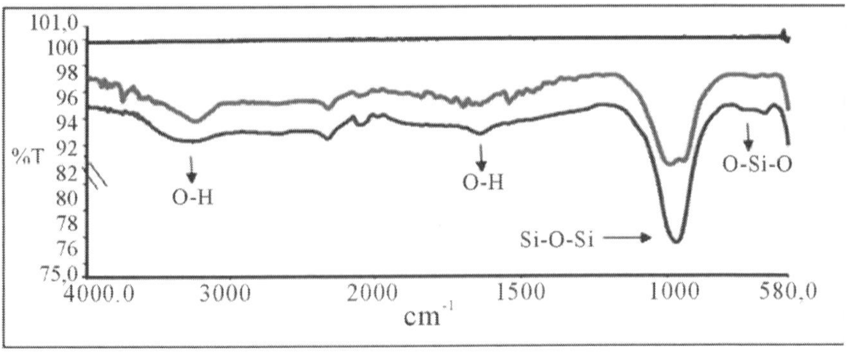

Figure 5. Overlapping spectra obtained by IR-ATR. Blue line: Na-LTA. Red line: Ca-LTA. Black line: the base of ATR.

Figure 6. SEM of the zynthesized zeolites. A: Na-LTA. B: Na-FAU.

3.4. Adsorption Isotherms

Several models can be used to describe the variation of adsorption data with metal ion concentration. The Langmuir isotherm [37] based on Equation (1) and Scatchard [38] graph based on Equation (2), where q_m indicates theoretical maximum sorption capacity of sorbent for a target solute to form a complete monolayer (also known as theoretical saturation capacity) and b is a constant related to affinity between adsorbent and adsorbate, does not fit very well with the experimental data.

$$\frac{1}{q} = \frac{1}{q_m} + \frac{1}{q_m b C_e} \tag{1}$$

$$\frac{q}{C_e} = q_m b - qb \tag{2}$$

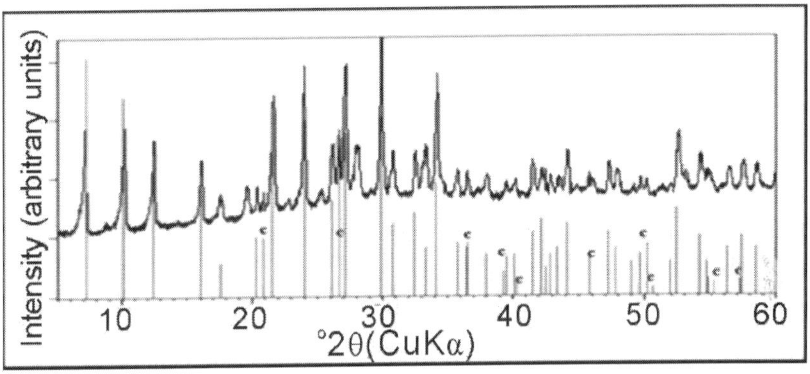

(A)

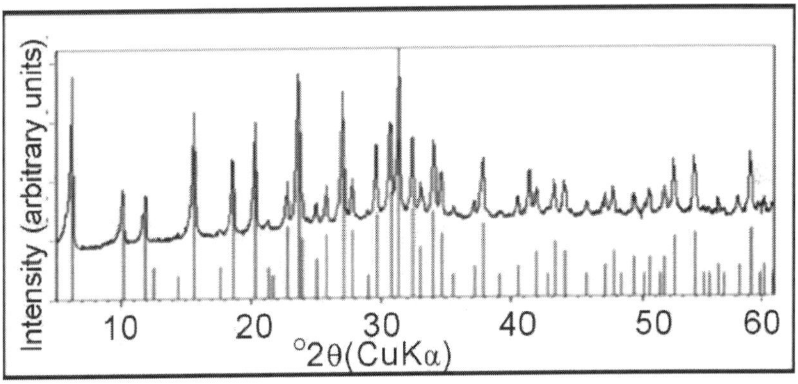

(B)

Figure 7. X-ray powder diffraction pattern. A: Na-LTA, where C is quartz. B: Na-FAU.

Therefore, the model employed was the Freundlich isotherm equation for dilute solutions given by Equation (3) [38], where k and m = (1/n) are the adsorption isotherm parameters, which are described as an indicator of sorption capacity and a constant related to sorption intensity respectively. The calculated parameters from the linearized form logq versus logC_e plot of the Freundlich model are collectively listed in Table 3.

$$q = kC_e^m \tag{3}$$

The Freundlich model seems to be agreeable for the zeolites in study due to the good fit that presents with the experimental data and R^2 values calculated from the plot [38]; hence, it can be deduced that the sorption of metals (II) on zeolites is a favorable process for each studied experimental conditions, due to the good values of R^2 as can seen in Table 3. The infinite surface coverage is predicted mathematically and indicates a multilayer sorption of the surface.

The sorption free energy has been evaluated to get some information of the nature of bonding (i.e. either physiosorption or chemisorption). In this regard the Dubinin-Radushkevich (D-R) isotherm [39] was tested as it seems in Equation (4).

$$\ln q = \ln q_m - K\varepsilon^2 \tag{4}$$

$$\varepsilon = RT\ln\left(1 + \frac{1}{C_e}\right) \tag{5}$$

The parameters K and ε are the activity coefficient in $mol^2 \cdot kJ^{-2}$ and the Polanyi potential, while R is the gas constant ($8.3143\ J \cdot mol^{-1} \cdot K^{-1}$) and T is the temperature in Kelvin. The activity coefficient constant is related with the sorption free energy as it seems in Equation (6).

$$E = \frac{1}{(-2k)^{1/2}} \tag{6}$$

Table 1. Instrumental and experimental conditions.

System	Parameter	Lead[a]	Cadmium[b]	Nickel[c]	Cobalt[d]
Flame (AAS)	Wavelength (nm)	217	228.8	232	240.7
	Current Lamp (mA)	4	4	4	7
	Slit (nm)	1	1	0.2	0.2
	Air/Acetylene ($L \cdot min^{-1}$)	2/10	2/10	2/10	2/10
Chemical	Sample and standard solutions pH*	2.5, 3.5⁺	1.5	2	2.5
	Complexing reagent, APDC (%w/v)	0.3	0.3	0.3	0.3
Flow Injection (22°C)	Sample flow rate ($mL \cdot min^{-1}$) for sample volume of 6 mL	2	4	5	4
	Complexing reagent flow rate, APDC ($mL \cdot min^{-1}$)	0.4	0.4	0.4	0.4
	Carrier flow rate, H_2O ($mL \cdot min^{-1}$)	4	4	4	3
	Eluent reagent volume, MIBK (μL)	100	200	175	100
Injector	Coil length (MR, cm)	170	100	300	150
	Preconcentration time (min)	3	1.5	1.2	1.5
	Elution time (s)	20	30	30	40
Mini-column	Length dimensions (cm) × inner diameter (mm)	0.3 × 4	0.3 × 4	0.3 × 4	0.3 × 4
	Quantity of material (mg)	20	20	20	20
Peristaltic Pump	Revolutions	15	15	15	15

*Adjusted with HNO_3. [a]$[Pb^{+2}] = 25\ \mu g \cdot L^{-1}$; [b]$[Cd^{+2}] = 40\ \mu g \cdot L^{-1}$; [c]$[Ni^{+2}] = 50\ \mu g \cdot L^{-1}$; [d]$[Co^{+2}] = 40\ \mu g \cdot L^{-1}$; ⁺Value just for Ca-LTA.

Table 2. Analytical parameters of the proposed method.

Metal	Zeolite	FIA sample introduction		Conventional sample introduction		LR ($\mu g \cdot L^{-1}$)	PF	SF	DL ($\mu g \cdot L^{-1}$)	RSD%
		Calibration curve	Regression factor (R^2)	Calibration curve	Regression factor (R^2)					
Pb^{+2}	Na-LTA	A = 0.0042x + 0.0234	0.9990	A = 0.0208y + 0.0064	0.9993	10 - 50	201.92	18	1.3	0.9
	Ca-LTA	A = 0.0041x + 0.0049	0.9982				197.12		1.7	1.4
	Na-FAU	A = 0.0039x + 0.0175	0.9996			10 - 100	187.50		0.3	0.8
	Ca-FAU	A = 0.0051x + 0.0054	0.9989				245.19		1.9	1.2
Cd^{+2}	Na-LTA	A = 0.0003x + 0.0198	0.9998	A = 0.0289y + 0.0093	0.9998	10 - 50	10.38	34	2.5	1.3
	Ca-LTA	A = 0.0007x + 0.0196	0.9992				24.22		5.6	1.6
	Na-FAU	A = 0.0009x + 0.0178	0.9987			30 - 60	31.14	17	5.3	1.0
	Ca-FAU	A = 0.0006x + 0.0165	0.9993				20.76		2.3	0.9
Ni^{+2}	Na-LTA	A = 0.0011x + 0.0057	0.9997	A = 0.0318y + 0.0076	0.9985	30 - 70	34.59	35	0.4	1.4
	Ca-LTA	A = 0.0012x + 0.0149	0.9994				37.74		0.5	0.3
	Na-FAU	A = 0.0025x + 0.0093	0.9998				78.62		0.7	0.2
	Ca-FAU	A = 0.0024x + 0.0151	0.9992				75.47		0.6	1.2
Co^{+2}	Na-LTA	A = 0.0022x + 0.0026	0.9996	A = 0.0120y + 0.0800	0.9992	0 - 50	183.30	28	1.1	1.9
	Ca-LTA	A = 0.0030x + 0.0068	0.9990				250.00		0.8	1.2
	Na-FAU	A = 0.0027x + 0.0052	0.9992				225.00	27	1.9	4.2
	Ca-FAU	A = 0.0028x + 0.0034	0.9991				233.33		2.1	3.4

A: absorbance; x: metal $\mu g \cdot L^{-1}$ concentration; y: metal $mg \cdot L^{-1}$ concentration; PF: preconcentration factor compared to the conventional introduction of aqueous sample. DL: detection limit. RSD %: repeatability. LR: lineal range. SF: sample frequency.

Table 3. Adsorption isotherm parameters of Freundlich.

Analyte	Zeolite	Equation	R^2	k	n
Lead	Na-FAU	$y = 2.871x - 0.504$	0.924	0.30	0.35
	Ca-FAU	$y = 1.945x + 1.041$	0.993	10.99	0.51
	Na-LTA	$y = 1.663x + 1.440$	0.989	27.51	0.60
	Ca-LTA	$y = 1.529x + 1.526$	0.979	33.57	0.65
Cadmium	Na-FAU	$y = 1.136x + 1.685$	0.998	48.41	0.88
	Ca-FAU	$y = 1.196x + 1.578$	0.998	37.84	0.84
	Na-LTA	$y = 2.510x + 0.783$	0.975	0.16	0.39
	Ca-LTA	$y = 2.500x + 0.767$	0.979	0.17	0.40
Nickel	Na-FAU	$y = 1.134x + 2.477$	0.998	299.90	0.88
	Ca-FAU	$y = 1.130x + 2.474$	0.996	297.90	0.88
	Na-LTA	$y = 1.101x + 2.196$	0.998	157.04	0.90
	Ca-LTA	$y = 1.134x + 2.137$	0.998	137.08	0.88
Cobalt	Na-FAU	$y = 1.042x + 2.709$	0.999	511.68	0.95
	Ca-FAU	$y = 1.045x + 2.710$	0.999	512.86	0.96
	Na-LTA	$y = 0.986x + 3.661$	0.999	4581.41	1.01
	Ca-LTA	$y = 0.988x + 3.701$	0.999	5023.42	1.01

The plot of $\ln q$ versus ε^2 is linear as show in Figure 8 for adsorption of $Co(PDC)_2$. The numerical values of different parameters are given in Table 4. The values of free energy of adsorption obtained for all zeolite samples lie in the range expected for physical adsorption (<8 kJ·mol^{-1}).

4. CONCLUSION

In the presented study a new solid phase extraction technique was developed based on the preconcentration on line of ions including Pb^{2+}, Ni^{2+}, Cd^{2+} and Co^{2+} ions in drinking water samples on synthetic zeolites type Na-LTA, Ca-LTA, Na-FAU and Ca-FAU and determination by FAAS. The conditions for quantitative and reproducible preconcentration, elution and FAAS determinations were studied. The system showed reproducibility and reliability in analytical data, with an R.S.D. value lower than 5% on fivefold experiment. 20 mg of zeolites can be used in more than 200 experiments without any loss in its sorption behavior. The system was successful in the preconcentration (20 < PF < 200) of analytes in volume sample of 6 mL. The sorption capacity under acidic conditions is a process based on interactions, between acid sites of zeolites and neutral chelates formed in the FIA separation and preconcentration system. The degrees of such interactions are highly independent because the systems are different for each element. However, the adsorptivity was high (k < 5020 Freundlich) and adsorption mechanism was via physisorption (E < 8 kJ·mol^{-1}). The method is simple, accurate, economical and environmentally friendly and can be applied for the determination of these understudy metal ions in environmental samples.

ACKNOWLEDGEMENTS

The authors are grateful to the FondoNacional de Ciencia y Tecnología (FONACIT) and Consejo de DesarrolloCientífico, Humanístico, Tecnológico y de lasArtes (CDCHTA) for financial support, Proyects

S1-97001192 and C-133105-08B respectively, Laboratorio de Análisis Químico Estructural de Materiales (LAQUEM) and Laboratorio de Cristalografía of the Universidad de Los Andes (ULA) Mérida-Venezuela.

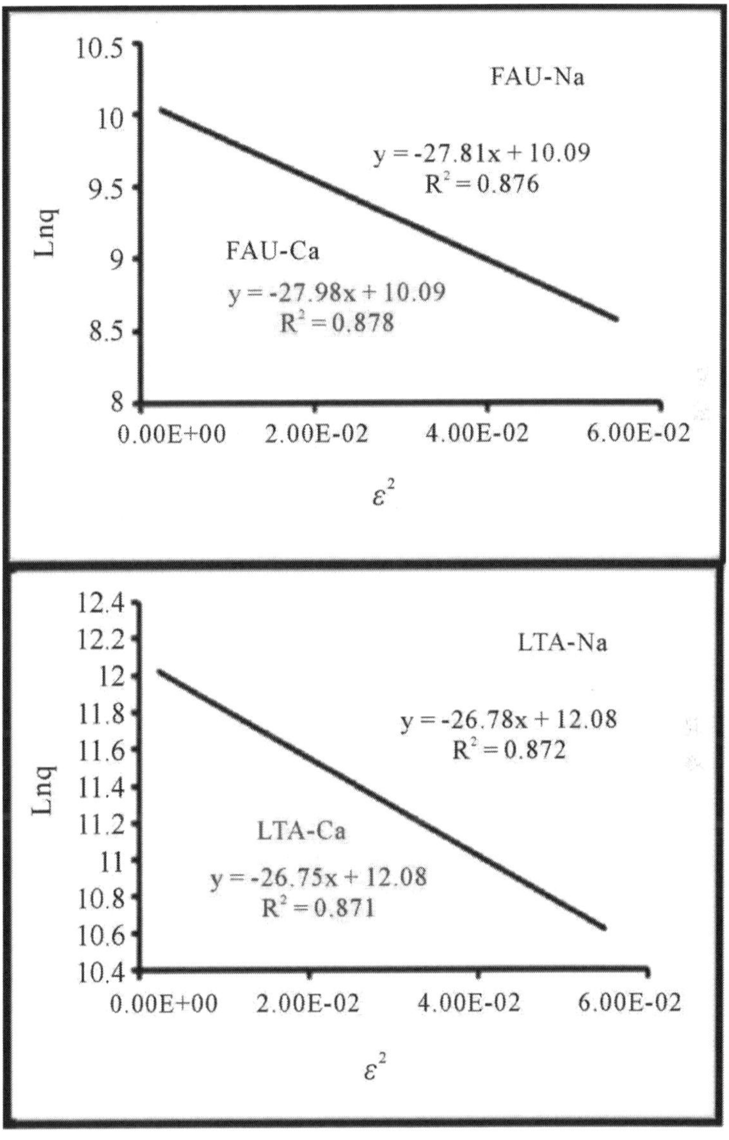

Figure 8. Isotherms D-R for the adsorption of $Co(PDC)_2$ on the synthetic zeolites.

Table 4. Adsorption isotherm parameters of Dubinin-Radushkevich.

Analyte	Zeolite	Equation	R^2	$-K$ (mol^2/kJ^2)	q_m ($mg \cdot g^{-1}$) × 10^3	E (kJ/mol)
Lead	Na-FAU	$y = -90.80x + 9.834$	0.985	90.80	18.65	0.074
	Ca-FAU	$y = -53.14x + 9.605$	0.843	53.14	14.84	0.097
	Na-LTA	$y = -46.20x + 9.502$	0.931	46.20	13.38	0.104
	Ca-LTA	$y = -39.64x + 9.159$	0.781	39.64	9.49	0.112
Cadmium	Na-FAU	$y = -147.8x + 8.689$	0.950	147.8	5.94	0.058
	Ca-FAU	$y = -160.5x + 8.702$	0.962	160.5	6.01	0.056
	Na-LTA	$y = -346.6x + 8.868$	0.967	346.6	7.10	0.038
	Ca-LTA	$y = -344.9x + 8.863$	0.973	344.9	7.07	0.038
Nickel	Na-FAU	$y = -173.1x + 10.62$	0.953	173.1	40.95	0.054
	Ca-FAU	$y = -173.6x + 10.63$	0.955	173.6	41.35	0.054
	Na-LTA	$y = -172.9x + 9.857$	0.953	172.9	19.09	0.054
	Ca-LTA	$y = -174.7x + 9.858$	0.953	174.7	19.11	0.054
Cobalt	Na-FAU	$y = -27.81x + 10.09$	0.876	27.81	24.10	0.134
	Ca-FAU	$y = -27.98x + 10.09$	0.878	27.98	24.10	0.134
	Na-LTA	$y = -26.78x + 12.08$	0.872	26.78	176.31	0.136
	Ca-LTA	$y = -26.75x + 12.08$	0.871	26.75	176.31	0.136

REFERENCES

1. M. R. Nabid, R. Sedghi, A. Bagheri, M. Behbahani, M. Taghizadeh, H. Abdi Oskooie and M. M. Heravi, "Preparation and Application of Poly(2-aminothiophenol)/ MWCNTs Nanocomposite for Adsorption and Separation of Cadmium and Lead Ions via Solid Phase Extraction," Journal of Hazardous Materials, Vol. 203-204, 2012, pp. 93-100.

2. G. Cheng, M. He, H. Peng and B. Hu, "Dithizone Modified Magnetic Nanoparticles for Fast and Selective Solid Phase Extraction of Trace Elements in Environmental and Biological Samples Prior to Their Determination by ICPOES," Talanta, Vol. 88, 2012, pp. 507-515.

3. Sánchez Trujillo, E. Vereda Alonso, A. García de Torres and J. M. Cano Pavón, "Development of a Solid Phase Extraction Method for the Multielement Determination of Trace Metals in Natural Waters Including Sea-Water by FI-ICP-MS," Microchemical Journal, Vol. 101, 2012, pp. 87-94.

4. M. Jamshidi, M. Ghaedi, K. Mortazavi, M. N. Biareh and M. Soylak, "Determination of Some Metal Ions by Flame- AAS after Their Preconcentration Using Sodium Dodecyl Sulfate Coated Alumina Modified with 2-hydroxy-(3- ((1-H-indol 3-yle)phenyl) methyl) 1-H-indol (2-HIYPMI)," Food and Chemical Toxicology, Vol. 49, No. 6, 2011, pp. 1229-1234.

5. P. S. Gonzáles, M. A. Firmino, C. S. Nomura, F. R. P. Rocha, P. V. Oliveira and I. Gaubeur, "Peat as a Natural Solid-Phase for Copper Preconcentration and Determination in a Multicommuted Flow System Coupled to Flame Atomic Absorption Spectrometry," Analytica Chimica Acta, Vol. 636, No. 2, 2009, pp. 198-204.

6. S. Samadi, H. Sereshti and Y. Assadi, "Ultra-Preconcentration and Determination of Thirteen Organophosphorus Pesticides in Water Samples Using Solid-Phase Extraction Followed by Dispersive Liquid-Liquid Microextraction and Gas Chromatography with Flame Photometric Detection," Journal of Chromatography A, Vol. 1219, 2012, pp. 61-65.

7. C. Duran, D. Ozdes, D. Sahin, V. N. Bulut, A. Gundogdu and M. Soylak, "Preconcentration of Cd(II) and Cu(II) Ions by

Coprecipitation without Any Carrier Element in Some Food and Water Samples," Microchemical Journal, Vol. 98, No. 2, 2011, pp. 317-322.

8. M. Chamsaz, A. Atarodi, M. Eftekhari, S. Asadpour and M. Adibi, "Vortex-Assisted Ionic Liquid Microextraction Coupled to Flame Atomic Absorption Spectrometry for Determination of Trace Levels of Cadmium in Real Samples," Journal of Advanced Research, Vol. 4, 2013, pp. 35-41.

9. S. Z. Mohammadi, H. Hamidian, L. Karimzadeh and Z. Moeinadini, "Tween 80 Coated Alumina: An Alternative Support for Solid Phase Extraction of Copper, Nickel, Cobalt and Cadmium Prior to Flame Atomic Absorption Spectrometric Determination," Arabian Journal of Chemistry, in Press, 2012.

10. R. Dobrowolski and M. Otto, "Determination of Nickel and Cobalt in Reference Plant Materials by Carbon Slurry Sampling GFAAS Technique after Their Simultaneous Preconcentration onto Modified Activated Carbon," Journal of Food Composition and Analysis, Vol. 26, No. 1-2, 2012, pp. 58-65.

11. S. Walas, A. Tobiasz, M. Gawin, B. Trzewik, M. Strojny and H. Mrowiec, "Application of a Metal Ion-Imprinted Polymer Based on Salen-Cu Complex to Flow Injection Preconcentration and FAAS Determination of Copper," Talanta, Vol. 76, 2008, pp. 96-101.

12. B. D. Koleva and E. Ivanova, "Flow Injection Analysis with Atomic Spectrometric Detection (Review Article)," Eurasian Journal of Analytical Chemistry, Vol. 3, No. 2, 2008, pp. 183-211.

13. L. Elci, A. A. Kartal and M. Soylak, "Solid Phase Extraction Method for the Determination of iron, Lead and Chromium by Atomic Absorption Spectrometry Using Amberite XAD-2000 Column in Various Water Samples," Journal of hazardous materials, Vol. 153, No. 1-2, 2008, pp. 454-461.

14. F. Marahel, M. Ghaedi, M. Montazerozohori, M. Nejati Biyareh, S. Nasiri Kokhdan and M. Soylak, "Solid-Phase Extraction and Determination of Trace Amount of Some Metal Ions on Duolite XAD 761 Modified with a New Schiff Base as Chelating Agent in Some Food Samples," Food and Chemical Toxicology, Vol. 49, 2011, pp. 208- 214.

15. H.-T. Fan, J. Li, Z.-C. Li and T. Sun, "An Ion-Imprinted Amino-Functionalized Silica Gel Sorbent Prepared by Hydrothermal Assisted Surface Imprinting Technique for Selective Removal of Cadmium (II) from Aqueous Solution," Applied Surface Science, Vol. 258, No. 8, 2012, pp. 3815-3822.

16. Bartyzel and E. M. Cukrowska, "Solid Phase Extraction Method for the Separation and Determination of Chromium(III) in the Presence of Chromium(VI) Using Silica Gel Modified by N,N-bis-(α-methylsalicylidene)- 2,2-dimethyl-1,3-propanediimine," Analytica Chimica Acta, Vol. 707, No. 1-2, 2011, pp. 204-209.

17. J. N. Bianchin, E. Martendal, R. Mior, V. N. Alves, C. S. T. Araújo, N. M. M. Coelho and E. Carasek, "Development of a Flow System for the Determination of Cadmium in Fuel Alcohol Using Vermicompost as Biosorbent and Flame Atomic Absorption Spectrometry," Talanta, Vol. 78, No. 2, 2009, pp. 333-336.

18. M. Ahmad, A. R. A. Usman, S. S. Lee, S.-C. Kim, J.-H. Joo, J. E. Yang and Y. S. Ok, "Eggshell and Coral Wastes as Low Cost Sorbents for the Removal of Pb^{2+}, Cd^{2+} and Cu^{2+} from Aqueous Solutions," Journal of Industrial and Engineering Chemistry, Vol. 18, 2012, pp. 198-204.

19. Tobiasz, S. Walas, A. Soto Hernández and H. Mrowiec, "Application of Multiwall Carbon Nanotubes Impregnated with 5-Dodecylsalicylaldoxime for On-Line Copper Preconcentration and Determination in Water Samples by Flame Atomic Absorption Spectrometry," Talanta, Vol. 96, 2012, pp. 89-95.

20. R. S. Amais, J. S. Ribeiro, M. G. Segatelli, I. V. P. Yoshida, P. O. Luccas and C. R. T. Tarley, "Assessment of Nanocomposite Alumina Supported on Multi-Wall Carbon Nanotubes as Sorbent for On-Line Nickel Preconcentration in Water Samples," Separation and Purification Technology, Vol. 58, 2007, pp. 122-128.

21. Y. P. d. Peña, W. López, J. L. Burguera, M. Burguera, M. Gallignani, R. Brunetto, P. Carrero, C. Rondon and F. Imbert, "Synthetic Zeolites as Sorbent Material for On- Line Preconcentration of Copper Traces and Its Determination Using Flame Atomic Absorption Spectrometry," Analytica Chimica Acta, Vol. 403, No. 1-2, 2000, pp. 249-258.

22. D. Afzali, A. Mostafavi, M. A. Taher and A. Moradian, "Flame Atomic Absorption Spectrometry Determination of Trace Amounts of Copper after Separation and Preconcentration onto TDMBAC-Treated Analcime Pyrocatechol-Immobilized," Talanta, Vol. 71, No. 2, 2007, pp. 971-975.

23. V. M. Nurchi and I. Villaescusa, "Sorption of Toxic Metal Ions by Solid Sorbents: A Predictive Speciation Approach Based on Complex Formation Constants in Aqueous Solution," Coordination Chemistry Reviews, Vol. 256, No. 1-2, 2012, pp. 212-221.

24. O. Martins, E. L. da Silva, E. Carasek, N. S. Gonçalves, M. C. M. Laranjeira and V. T. de Fávere, "Chelating Resin from Functionalization of Chitosan with Complexing Agent 8-Hydroxyquinoline: Application for Metal Ions on Line Preconcentration System," Analytica Chimica Acta, Vol. 521, No. 2, 2004, pp. 157-162.

25. D. Afzali, A. Mostafavi, F. Etemadi and A. Ghazizadeh, "Application of Modified Multiwalled Carbon Nanotubes as Solid Sorbent for Separation and Preconcentration of Trace Amounts of Manganese Ions," Arabian Journal of Chemistry, Vol. 5, No. 2, 2012, pp. 187-191.

26. N. Anthemidis, G. Giakisikli, S. Xidia and M. Miró, "On-Line Sorptive Preconcentration Platform Incorporating a Readily Exchangeable Oasis HLB Extraction Micro-Cartridge for Trace Cadmium and Lead Determination by Flow Injection-Flame Atomic Absorption Spectrometry," Microchemical Journal, Vol. 98, 2011, pp. 66- 71.

27. Y. Petit de Peña, B. Paredes, W. Rondón, M. Burguera, J. L. Burguera, C. Rondón, P. Carrero and T. Capote, "Continuous Flow System for Lead Determination by Faas in Spirituous Beverages with Solid Phase Extraction and On-Line Copper Removal," Talanta, Vol. 64, No. 5, 2004, pp. 1351-1358.

28. C. Baerlocher and L. B. McCusker, "Database of Zeolite Structures," 2013. http://www.iza-structure.org/databases

29. M. Al-Anber and Z. A. Al-Anber, "Utilization of Natural zeolite as Ion-Exchange and Sorbent Material in the Removal of Iron," Desalination, Vol. 225, No. 1-3, 2008, pp. 70-81.

30. O. L. Corona, M. A. Hernández, F. Hernández, F. Rojas, R. Portillo, V. H. Lara and F. M. Carlos, "Propiedades de Adsorción en Zeolitas con Anillos de 8 Miembros: I. Microporosidad y Superficie Externa," Matéria (Río J.), Vol. 14, No. 3, 2009, pp. 918-931.

31. C. R. Melo, H. G. Riella, N. C. Kuhnen, E. Angioletto, A. R. Melo, A. M. Bernardin, M. R. da Rocha and L. da Silva, "Synthesis of 4A Zeolites from Kaolin for Obtaining 5A Zeolites through Ionic Exchange for Adsorption of Arsenic," Materials Science and Engineering B, Vol. 177, No. 4, 2012, pp. 345-349.

32. F. E. Imbert, C. Moreno, A. Montero, B. Fontal and J. Lujano, "Venezuelan Natural Alumosilicates as a Feedstock in the Synthesis of Zeolite A," Zeolites, Vol. 14, No. 5, 1994, pp. 374-378.

33. JCPDS, "PDF-2 Database ICDD," Newton Square, 2001.

34. Y. M. Liew, H. Kamarudin, A. M. Mustafa Al Bakri, M. Luqman, I. Khairul Nizar, C. M. Ruzaidi and C. Y. Heah, "Processing and Characterization of Calcined Kaolin Cement Powder," Construction and Building Materials, Vol. 30, 2012, pp. 794-802.

35. R. T. Rigo, S. B. C. Pergher, D. I. Petkowicz and J. H. Z. d. Santos, "A New Procedure for a Zeolite Synthesis from Natural Clays," Química Nova, Vol. 32, No. 2009, pp. 21-25.

36. T. Frising and P. Leflaive, "Extraframework Cation Distributions in X and Y Faujasite Zeolites: A Review," Microporous and Mesoporous Materials, Vol. 114, No. 1-3, 2008, pp. 27-63.

37. M. Tuzen, K. Parlar and M. Soylak, "Enrichment/Separation of Cadmium(II) and Lead(II) in Environmental Samples by Solid Phase Extraction," Journal of Hazardous Materials, Vol. 121, No. 1-3, 2005, pp. 79-87.

38. O. Gezici, H. Kara, A. Ayar and M. Topkafa, "Sorption Behavior of Cu(II) Ions on Insolubilized Humic Acid under Acidic Conditions: An Application of Scatchard Plot Analysis in Evaluating the pH Dependence of Specific and Nonspecific Bindings," Separation and Purification Technology, Vol. 55, 2007, pp. 132-139.

39. O. Gezici, H. Kara, M. Ersoz and Y. Abali, "The Sorption Behavior of a Nickel-Insolubilized Humic Acid System in a Column Arrangement," Journal of Colloid and Interface Science, Vol. 292, No. 2, 2005, pp. 381-391.

CHAPTER 8

Determination of Heavy Metals at Sub-ppb Levels in Water by Graphite Furnace Atomic Absorption Spectrometry Using a Direct Introduction Technique after Preconcentration with an Iminodiacetate Extraction Disk

Tetsuo Inui*, Atsuko Kosuge, Atsushi Ohbuchi, Kazuhiro Fujita, Yuya Koike, Masaru Kitano, Toshihiro Nakamura

Department of Applied Chemistry, Meiji University, Kawasaki, Japan

ABSTRACT

A direct analysis method combining an iminodiacetate extraction disk (IED) with graphite furnace atomic absorption spectrometry was developed for the determination of Co, Ni, Cu, Cd, Sn, Pb, and Bi at sub-ppb levels in water. A 100 mL water sample was adjusted to pH 5.6 with nitric acid and a 1 mol·L^{-1} ammonium acetate solution, and then passed through an IED (diameter, 47 mm; effective filtering diameter, 35 mm) at a flow rate of 80 - 100 mL·min^{-1} to preconcentrate seven analytes. The IED was dried at 100°C for 20 min in an electric oven, and 110 - 145 small disks, each 2 mm in diameter, were punched out from the IED. A small disk was introduced into the graphite furnace and atomized according to a heating program. For Cd, Sn, Pb, and Bi measurements, Pd was used as a chemical modifier to enhance the absorbances. Calibration was performed using aqueous standard solutions. The detection limits, corresponding to three times the standard deviation (n = 5) of the blank values, were 0.092 $mg{\cdot}L^{-1}$ for Co, 0.12 $mg{\cdot}L^{-1}$ for Ni, 0.40 $mg{\cdot}L^{-1}$ for Cu, 0.077 $mg{\cdot}L^{-1}$ for Cd, 0.92 $mg{\cdot}L^{-1}$ for Sn, 0.61 $mg{\cdot}L^{-1}$ for Pb, and 0.80 $mg{\cdot}L^{-1}$ for Bi with an enrichment factor of 140 using a 100-mL water sample. A spike test for the seven analytes in tap water, rainwater, river water, and mineral drinking water showed quantitative recoveries (93% - 108%).

KEYWORDS

Water Sample; Heavy Metal; Preconcentration; Iminodiacetate Extraction Disk; Direct Introduction Technique; Graphite Furnace Atomic Absorption Spectrometry

1. INTRODUCTION

Some heavy metals are considered to be essential trace elements for the human body because they act as catalysts in enzymatic reactions and are deeply involved in the metabolism of glucose and lipids. In contrast, other heavy metals have higher toxicity and carcinogenicity. Therefore, it is important to determine the heavy metal content of environmental samples.

Instrumental sensitivity is not always sufficient for the determination of heavy metals in water because their concentrations are generally at sub-ppb levels. Therefore, it is necessary to combine a preconcentration technique with an instrumental analysis method, such as atomic absorption spectrometry (AAS), inductively coupled plasma atomic emission spectrometry (ICP-AES), and inductively coupled plasma mass spectrometry (ICP-MS), for the determination of heavy metals in water.

Solid-phase extraction (SPE) has recently become popular for separation and/or preconcentration [1]. In particular, disk-type SPE has been widely used to preconcentrate a large-volume water sample (e.g., 100 - 1000 mL) for inorganic [2-6] and organic [7-11] analyses because it offers higher flow rates than column-type SPE owing to its large cross-sectional area. Iminodiacetate Extraction Disks (IEDs) are useful for the preconcentration of trace metals in water because of their high selectivity for heavy metals. Kuriyama et al. [12] and Ouyang et al. [13] reported a method for the determination of Al, Mn, Fe, Co, Ni, Cu, Zn, Cd, and Pb in water by ICP-AES with an IED. A 1000-mL water sample (pH 5.6 - 5.8) was passed through an IED at a flow rate of 100 - 150 $mL \cdot min^{-1}$ to preconcentrate the nine analytes, and these analytes were eluted from the IED with 9 - 10 mL of 3 $mol \cdot L^{-1}$ nitric acid. The eluate was then diluted to 10 - 20 mL with purified water or distilled water. Lee et al. [14] also reported a method for the determination of V, MnCo, Ni, Cu, Mo, Cd, and U in water by ICP-MS with an IED. However, the elution step may give rise to a negative error owing to incomplete elution of the analytes from the IED. Therefore, direct analysis of analytes in the IED is necessary to avoid analyte loss.

Several direct analysis methods combining an IED with Laser-Induced Breakdown Spectroscopy (LIBS) [15, 16] or X-ray fluorescence (XRF) spectrometry [17,18] have been reported for the determination of heavy metals in water. Direct introduction/AAS [19] is also attractive. However, a method combining IED with direct introduction/AAS has not yet been developed for the determination of heavy metals in water. Furthermore, it is difficult to introduce an IED directly into the furnace because a commercially available IED is 47 mm in diameter. This problem can be solved by adopting a punching technique before AAS analysis [20-23].

The present investigation concerns a further application of direct introduction/AAS to determine heavy metals at sub-ppb levels in water. This method involves preconcentration of Co, Ni, Cu, Cd, Sn, Pb, and Bi with an IED. Furthermore, the discussion is extended to elemental mapping, the size of the small disk, the atomization conditions, and the calibration curves and detection limits. The detection limits of the proposed method are also compared with those reported in the literature [12-15, 17,18]. The proposed method was successfully applied for the determination of the seven analytes in tap water, rainwater, river water, and mineral drinking water.

2. EXPERIMENTAL

2.1. Apparatus

A ZEEnit 60 atomic absorption spectrometer (Analytik Jena AG, Jena, Germany) was used in conjunction with Hitachi hollow-cathode lamps, a transversely heated graphite atomizer, and an SSA 61Z solid sampling system (Analytik Jena AG). The Zeeman effect was employed for background correction under two-field mode (0.8 T magnetic field strength) conditions. Pure argon (99.995%) was used as the carrier gas during drying and pyrolysis, except during atomization, where the gas flow was stopped. Atomization was conducted in Solid Sampling (SS) graphite tubes (Analytik Jena AG, Part No. 407-152.013) with SS graphite platforms (Analytik Jena AG, Part No. 407-152.023). The integrated absorbance (peak area) was used exclusively for signal evaluation and quantification. The instrumental conditions are given in Table 1.

A micro energy-dispersive XRF (μ-EDXRF) spectrometer (Rigaku Corp., Tokyo, Japan) was used for elemental mapping of the IED to confirm the homogeneities of the analytes on it.

Aqueous samples were injected into the SS platform using an Eppendorf Model Reference—4800 pipette (10 μL fixed volume, Eppendorf, Hamburg, Germany).

A Demi-Ace Model-15 deionizer (Kurita Water Industries, Tokyo, Japan) was used to prepare deionized water. An HM-30V pH meter (DKK-TOA, Tokyo, Japan) was used for pH control of the sample solution. An FS-320 electric oven (Advantec Toyo Kaisha, Tokyo, Japan) was used to dry the IED.

2.2. Reagents and Samples

Stock standard solutions (1000 mg·L^{-1}) of Co, Ni, Cu, Cd, Sn, Pb, and Bi were purchased from Junsei Chemical (Tokyo, Japan). Calibrating standard solutions of Co, Ni, Cu, Cd, Sn, Pb, and Bi were prepared by accurate dilution of the respective stock standard solutions. A 1000 mg·L^{-1} Pd stock standard solution for AAS (analyticalreagent grade, Wako Pure Chemical, Osaka, Japan) was used as a chemical modifier.

An IED (3M Empore™ extraction disk, 47 mm diameter, 0.5 mm thickness, 3M, St. Paul, MN) was used throughout the study.

Tap water samples were obtained from Kawasaki and Fujisawa, Kanagawa, Japan. Rainwater was sampled at Kawasaki, Kanagawa, Japan. River water was sampled from the Tamagawa River, Okutama, Tokyo, Japan. Bottled mineral drinking water samples were brought from Kawasaki, Kanagawa, Japan. Water samples were filtered through a glass-fiber filter (GMF 150, 1.0 μm pore size, 47 mm diameter, 0.73 mm thickness, Whatman, Kent, England) and analyzed immediately using the proposed method.

2.3. Procedure

An IED (diameter, 47 mm; effective filtering diameter, 35 mm) was soaked with methanol and then conditioned successively with 20 mL of 3 mol·L^{-1} nitric acid, 100 mL of deionized water, and 50 mL of 0.1 mol·L^{-1} ammonium acetate solution (pH 5.6). A 100-mL water sample was first adjusted to pH 5.6 [12-14,18] with nitric acid and a 1 mol·L^{-1} ammonium acetate solution because the carboxylate groups in the iminodiacetic acid are negatively charged at a pH of greater than 5, and the IED functions as a cation exchanger or metal chelator when the pH approaches neutral. After pH adjustment, the 100-mL water sample was passed through the IED at a flow rate of 80 - 100 mL·min^{-1} [18] to preconcentrate Co, Ni, Cu, Cd, Sn, Pb, and Bi. The IED was subsequently washed with 20 mL of deionized water. It was then dried at 100°C for 20 min in the electric oven, and 110 - 145 small disks, each 2 mm in diameter, were punched out from the IED.

Table 1. Instrumental conditions of atomic absorption spectrometry for Co, Ni, Cu, Cd, Sn, Pb, and Bi.

	Co	Ni	Cu	Cd	Sn	Pb	Bi
Analytical line/nm	240.7	232.0	324.8	228.8	224.6	283.3	223.1
Lamp current/mA	5	3	5	4	8	7	7
Slit width/nm	0.2	0.2	0.8	1.2	0.8	0.8	0.2

A small disk was placed on the SS platform with a wooden toothpick and introduced into the graphite furnace. For the Cd, Sn, Pb, and Bi measurements, suitable amounts of Pd were added before introduction into the furnace. The specimen was dried, pyrolyzed, and atomized according to the heating program described below (Table 2). Calibration was performed using the aqueous standards, which were pipetted manually onto the SS platform, introduced into the graphite furnace, and analyzed under the same atomization conditions. Calibration curves for the seven analytes were built by plotting the integrated absorbance versus the amount. The analyte concentrations in water were calculated using the following equation:

$$C = A\left(S_{\mathrm{IED}}/S_{\mathrm{small\,disk}}\right)/V \tag{1}$$

where C ($mg\cdot L^{-1}$) is the analyte concentration in water, A (ng) is the absolute amount of analyte on the small disk obtained from the calibration curve, S_{IED} is the effective filtration area of the IED (962 mm^2), $S_{small\ disk}$ is the area of the small disk (3.14 mm^2), and V (mL) is the sample volume.

3. RESULTS AND DISCUSSION

3.1. Elemental Mapping

The analytes on the IED should have a homogeneous distribution because the analyte concentrations in water are based on the area ratio of the IED and the small disk (Equation (1)). The homogeneities of the analytes on the IED were evaluated using elemental mapping of the fluorescent X-ray intensities. For XRF measurements, it is necessary to preconcentrate the analytes to a high density on the IED before they break through. In this experiment, therefore, a 100-mL deionized water sample (pH 5.6) containing 1 mg each of Co, Ni, Cu, Cd, Sn, Pb, and Bi was passed through the conditioned IED at a flow rate of 1 $mL\cdot min^{-1}$. The IED was then dried at 100°C for 20 min in the electric oven, coated with a commercially available laminate film to prevent damage to the

IED by X-ray irradiation [6,18], and measured using m-EDXRF spectrometer. The mapping conditions were 100 points (1 mm diameter) in 18 ´ 18 mm^2 of the scanning area. As an example, elemental mapping images of Cu, Cd, and Pb on the IED are shown in Figure 1. Each X-ray intensity varied slightly, and the relative standard deviations (RSDs) (n = 100) of the X-ray intensities were 1.8% for Co Kα, 2.1% for Ni Kα, 2.0% for Cu Kα, 3.3% for Cd Kα, 4.1% for Sn Kα, 2.1% for Pb Lβ, and 3.4% for Bi Lα. The results (i.e., RSDs of less than 5%) indicate that the analytes are homogeneously distributed on the IED. Consequently, the analyte concentrations in water can be determined using Equation (1).

3.2. Size of Small Disks

We next investigated the size of the small disks from the IED introduced into the furnace. When the IED was punched with three different sizes of hole punch (1, 2, and 3 mm diameter), the sample masses were 0.26, 1.20, and 3.18 mg, respectively. Considering the sample mass (typically 1 mg) in direct introduction/AAS [19], we decided to use the 2-mm-diameter hole punch for this study. In addition, 372 small disks, each 2 mm in diameter, were punched out from the IED and then weighed directly on the SS platforms. The average mass of the small disks was 1.29 mg, and the RSD (n = 372) of the mass was 7.4%. In fact, it is possible to conduct repeated measurements of the analytes because 110 - 145 small disks, each 2 mm in diameter, could be prepared from the effective filtration area of an IED. Furthermore, by preserving small disks containing known amounts of the analytes in a desiccator, equipment calibration can also be performed using them.

3.3. Atomization Conditions

Our laboratory has successfully applied the direct introduction/AAS method to a variety of inorganic materials, such as carbonate rocks [24], silicate rocks [25], quartz and rock crystals [26], zirconia [27], silicon nitride and silicon carbide [28], human and artificial teeth [29], and soils

[30]. If the pyrolysis temperature is insufficient for the decomposition of an IED comprising 90% poly(styrenedivinylbenzene) copolymer functionalized with an iminodiacetic acid group and 10% polytetrafluoroethylene, extremely high background absorption caused by the evolution of smoke may be observed during the atomization step. Therefore, a pyrolysis temperature of 800°C (60 s) was adopted to remove the organic matrix before atomization. In contrast, the analytes were atomized at the temperatures recommended by the manufacturer. When the analytes (Co, Ni, Cu, Cd, Sn, Pb, and Bi) on a small disk (2 mm diameter) from the IED were atomized under the above-mentioned conditions, the absorbancetime profiles of only Pb and Bi showed double peaks consisting of a large first peak and a small second peak, as shown in Figure 2(a). In contrast, when a palladium nitrate solution was added to the small disk as a chemical modifier for the elements of the 13 - 16 groups [31] before it was introduced into the furnace, the absorbance-time profiles of Pb and Bi showed a single broad peak, as shown in Figure 2(b). The results suggest that the first peak may be due to vaporization of the analyte element, and the second peak may be attributable to condensation and revaporization of the analyte element. The reason is that the double peaks could be eliminated by the addition of Pd, which prevents the evaporation of elements of the 13 - 16 groups by forming a thermally stable compound (i.e., a solid solution and/or intermetallic compound). A comparison of the absorbance-time profiles of Pb and Bi reveals that the Bi signal was almost the same as the background signal. The overlap of the background signal with the Bi signal did not become a serious problem because background correction was carried out using the Zeeman effect instead of continuum sources (mainly deuterium lamp) throughout the study [32]. In addition, Cd, Sn, Pb, and Bi may evaporate during pyrolysis because these analytes are volatile. Palladium is also effective in preventing evaporation of Cd [33,34]. Therefore, the effect of the amount of Pd on the integrated absorbances of Cd, Sn, Pb, and Bi was investigated. Figure 3 shows the variation in the integrated absorbances of Cd, Sn, Pb, and Bi on the small disk with the amount of Pd. The integrated absorbances of Sn, Pb, and Bi increased with an increase in the amount of Pd and leveled off above 0.25 mg for Pb and 2 μg for Sn and Bi. In contrast, the integrated absorbance of Cd increased remarkably up to 4 μg.

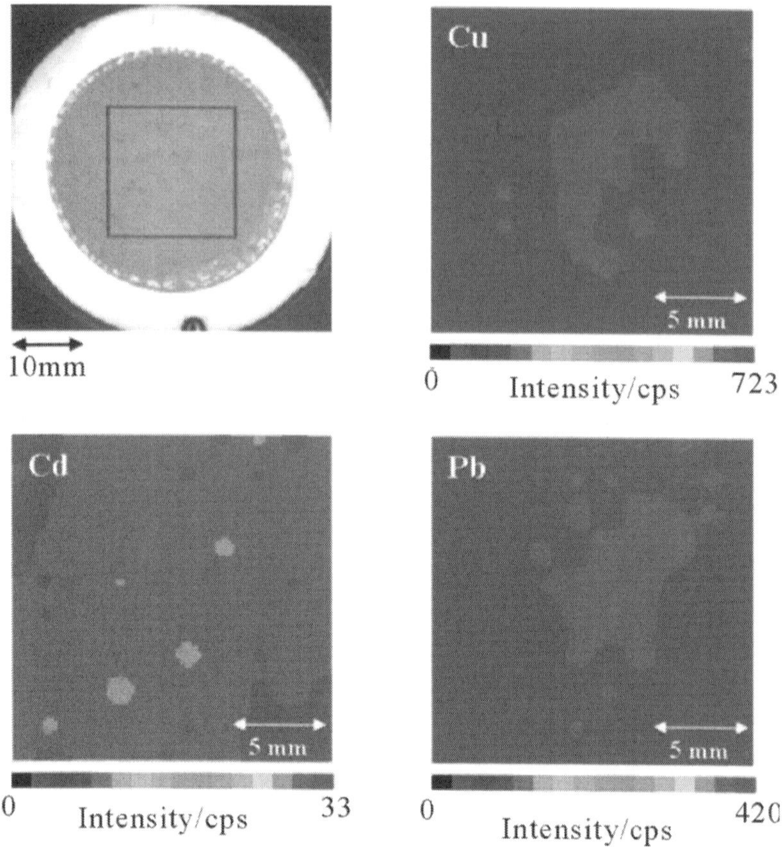

Figure 1. Elemental mapping images of Cu, Cd, and Pb on the IED.

The atomization conditions for the determination of Co, Ni, Cu, Cd, Sn, Pb, and Bi are given in Table 2.

3.4. Calibration Curves and Detection Limits

In direct introduction/AAS, calibration is performed using solid standards or aqueous standards [19]. Because solid standards are more expensive, it is preferable to use aqueous standards for calibration. However, the difference in matrices between the solid and liquid forms may cause different atomization processes. Therefore, we prepared calibration curves (Figure 4) for Co, Ni, Cu, Cd, Sn, Pb, and Bi by using two series of calibrating standards consisting of the analytes in a small disk (2 mm diameter) from the IED and an aqueous standard under the proposed atomization conditions (Table 2). The calibration curves established using both standards showed good linearity at less than 3 ng for Co, Ni, and Cu; 0.5 ng for Cd; 10 ng for Sn and Bi; and 20 ng for Pb. A decrease in the integrated absorbance was also observed for all the analytes except for Pb as a result of the Lorentz pressure effect. Furthermore, the slopes of the calibration curves for the small disk were almost identical to those for the aqueous standard, indicating that the aqueous standard can be used to determine the seven analytes on the small disk. The detection limits, corresponding to three times the standard deviation (n = 5) of the blank values, were 0.030 ng for Co, 0.038 ng for Ni, 0.13 ng for Cu, 0.025 ng for Cd, 0.30 ng for Sn, 0.20 ng for Pb, and 0.26 ng for Bi. When an enrichment factor of 140, calculated by dividing the sample volume (ca. 100 g) by the mass of an IED (ca. 0.7 g), was achieved using a 100-mL water sample, the detection limits were 0.092 $mg \cdot L^{-1}$ for Co, 0.12 $mg \cdot L^{-1}$ for Ni, 0.40 $mg \cdot L^{-1}$ for Cu, 0.077 $mg \cdot L^{-1}$ for Cd, 0.92 $mg \cdot L^{-1}$ for Sn, 0.61 $mg \cdot L^{-1}$ for Pb, and 0.80 $mg \cdot L^{-1}$ for Bi.

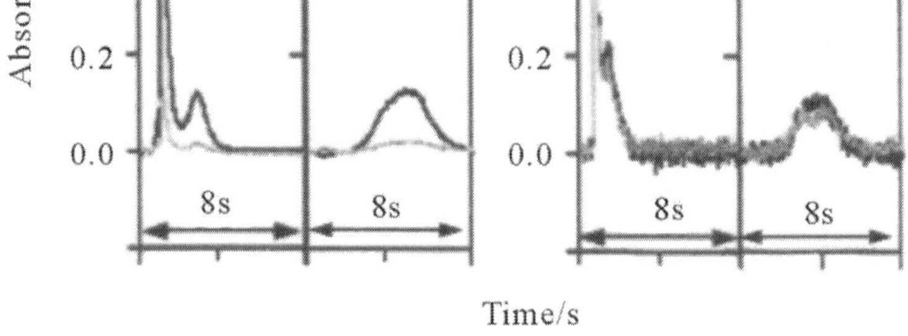

Figure 2. Absorbance-time profiles of Pb and Bi on a small disk (2 mm diameter) from the IED (a) without and (b) with the addition of $Pd(NO_3)_2$ solution. Black line is the atomic absorption; gray line is the background absorption.

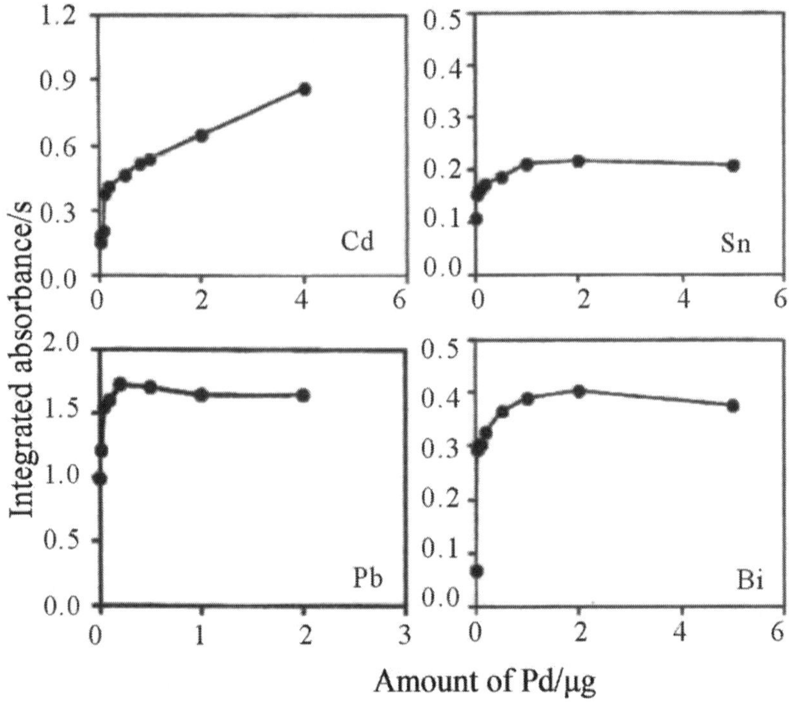

Figure 3. Variation in the integrated absorbances of Cd, Sn, Pb, and Bi on a small disk (2 mm diameter) from the IED with the amount of Pd.

Table 3 compares the detection limits of the proposed method with those obtained by combining an IED with ICP-AES [12,13], ICP-MS [14], LIBS [15], or XRF spectrometry [17,18] for the determination of heavy metals in water. The detection limits of the proposed method are lower than those of LIBS and comparable to those of XRF spectrometry. However, although the detection limits are higher than those of ICP-AES and ICP-MS, which involve an elution step, at least sub-ppb levels of heavy metals in water can be determined directly without an elution step by using the proposed method.

Table 2. Atomization conditions for the determination of Co, Ni, Cu, Cd, Sn, Pb, and Bi.

	Co	Ni	Cu	Cd	Sn	Pb	Bi
Drying/°C	0 - 120 (45)	0 - 120 (45)	0 - 120 (45)	0 - 120 (45)	0 - 120 (45)	0 - 120 (45)	0 - 120 (45)
Pyrolysis/°C	800 (60)	800 (60)	800 (60)	800 (60)	800 (60)	800 (60)	800 (60)
Atomization/°C	2400 (8)	2500 (8)	2300 (8)	1700 (8)	2700 (5)	1900 (8)	2200 (8)
Chemical modifier	-	-	-	4 μg Pd	2 μg Pd	0.25 μg Pd	2 μg Pd

Numbers in parentheses: heating time/s.

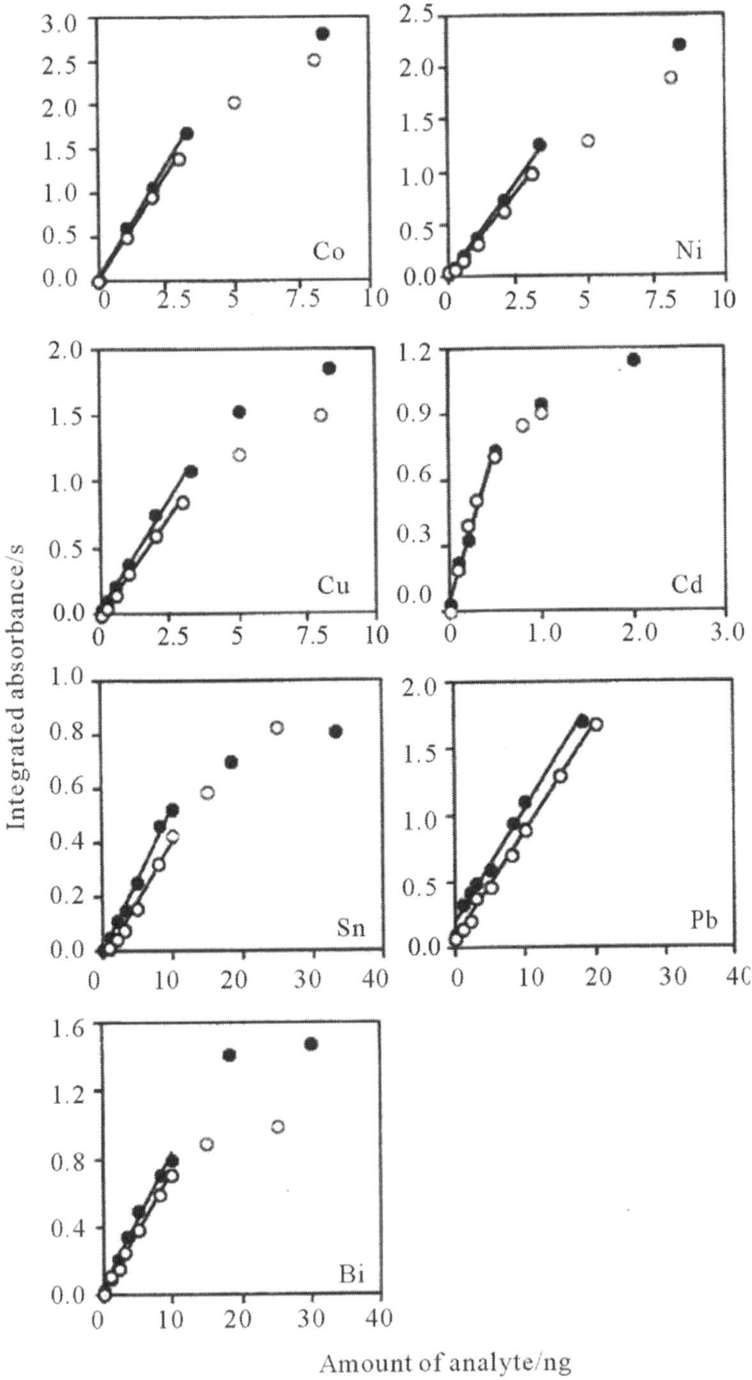

Figure 4. Calibration curves of Co, Ni, Cu, Cd, Sn, Pb, and Bi in a small disk (2 mm diameter) from the IED (●) and an aqueous standard (○).

3.5. Application to Water Samples

We first investigated the repeatability of the proposed method using 100 mL synthetic mixtures of 6 mg·L^{-1} Co, 9 mg·L^{-1} Ni, 9 mg·L^{-1} Cu, 0.9 mg·L^{-1} Cd, 15 mg·L^{-1} Sn, 15 mg·L^{-1} Pb, and 6 mg·L^{-1} Bi. The RSDs for five measurements were 4.4% for Co, 3.2% for Ni, 5.0% for Cu, 6.2% for Cd, 4.7% for Sn, 5.1% for Pb, and 4.8% for Bi. Next the proposed method was applied to the determination of Co, Ni, Cu, Cd, Sn, Pb, and Bi in tap water, rainwater, river water, and mineral drinking water. Calibrations were performed using a suite of standard solutions, and the analyte concentrations in the water samples were calculated using Equation (1). The accuracy of the analytical values was evaluated by a spike test for the seven analytes in 100- mL water samples. The results of the spike test are given in Table 4. The recoveries for the seven analytes in the water samples were satisfactory (93% - 108%). The good recoveries indicate that the proposed method is useful for the determination of heavy metals at sub-ppb levels in water.

4. CONCLUSION

A direct analysis method combining IED with graphite furnace AAS was proposed for the determination of Co, Ni, Cu, Cd, Sn, Pb, and Bi at sub-ppb levels in water. An IED was used to preconcentrate the seven analytes. The seven analytes could be determined without an elution step by directly introducing a small disk (2 mm diameter) from the IED into the furnace. The slopes of the calibration curves established using the small disk agreed well with those established using an aqueous standard, which indicated that calibration against aqueous standards is feasible. When an enrichment factor of 140 was obtained using a 100-mL water sample, the detection limits were 0.092 mg·L^{-1} for Co, 0.12 mg·L^{-1} for Ni, 0.40 mg·L^{-1} for Cu, 0.077 mg·L^{-1} for Cd, 0.92 mg·L^{-1} for Sn, 0.61 mg·L^{-1} for Pb, and 0.80 mg·L^{-1} for Bi. A spike test for the seven analytes in tap water, rainwater, river water, and mineral drinking water showed quantitative recoveries (93% - 108%).

Table 3. Comparison of the detection limit of the proposed method with those of the published techniques for the determination of heavy metals in water using IED preconcentration.

Technique	Sample volume (mL)	Enrichment factor	Detection limit ($\mu g \cdot L^{-1}$)							Reference
			Co	Ni	Cu	Cd	Sn	Pb	Bi	
ICP-AES	1000	100	0.020	0.064	0.098	0.003	-	0.058	-	12
ICP-AES	1000	50	0.003	0.005	0.002	0.001	-	0.011	-	13
ICP-MS	1	2	0.006	0.034	0.050	0.008	-	0.110	-	14
LIBS	10	-	-	310	4.2	210	-	1100	-	15
EDXRF	1000	>1600	-	0.6	0.4	3.8	-	0.3	-	17
WDXRF	1000	-	0.1	0.1	0.1	7.1	-	0.8	-	18
GFAAS	100	140	0.092	0.12	0.40	0.077	0.92	0.61	0.80	This work

ICP-AES: Inductively coupled plasma atomic emission spectrometry; ICP-MS: Inductively coupled plasma mass spectrometry; LIBS: Laser-induced breakdown spectroscopy; EDXRF: Energy dispersive X-ray fluorescence spectrometry; WDXRF: Wavelength dispersive X-ray fluorescence spectrometry; GFAAS: Graphite furnace atomic absorption spectrometry.

Table 4. Spike tests for Co, Ni, Cu, Cd, Sn, Pb, and Bi added to six water samples.

Element	Sample	Added (μg·L⁻¹)	Found (μg·L⁻¹)	Recovery (%)
Co	Tap water (Kawasaki)	0	0.86	
		3.0	4.0	103
	Tap water (Fujisawa)	0	<0.092*	
		3.0	3.0	100
	Rainwater (Kawasaki)	0	<0.092*	
		3.0	3.0	100
	River water (Okutama)	0	<0.092*	
		3.0	3.0	100
	Mineral drinking water (Brand A)	0	<0.092*	
		3.0	3.1	102
	Mineral drinking water (Brand B)	0	<0.092*	
		3.0	3.2	107
Ni	Tap water (Kawasaki)	0	0.20	
		3.0	3.1	95
	Tap water (Fujisawa)	0	0.40	
		3.0	3.4	99
	Rainwater (Kawasaki)	0	1.3	
		3.0	4.1	94
	River water (Okutama)	0	0.71	
		3.0	4.0	108
	Mineral drinking water (Brand A)	0	<0.12*	
		3.0	2.9	98
	Mineral drinking water (Brand B)	0	0.17	
		3.0	3.1	97
Cu	Tap water (Kawasaki)	0	2.8	
		3.0	6.0	106
	Tap water (Fujisawa)	0	4.2	
		3.0	7.2	100
	Rainwater (Kawasaki)	0	7.1	
		3.0	10	99
	River water (Okutama)	0	1.5	
		3.0	4.5	101
	Mineral drinking water (Brand A)	0	1.1	
		3.0	4.0	95
	Mineral drinking water (Brand B)	0	1.0	
		3.0	4.2	105
Cd	Tap water (Kawasaki)	0	<0.077*	
		1.2	1.2	97
	Tap water (Fujisawa)	0	<0.077*	
		1.2	1.1	94
	Rainwater (Kawasaki)	0	<0.077*	
		1.2	1.3	107
	River water (Okutama)	0	<0.077*	
		1.2	1.2	103
	Mineral drinking water (Brand A)	0	<0.077*	
		1.2	1.2	98
	Mineral drinking water (Brand B)	0	<0.077*	
		1.2	1.2	102

Pb	Tap water (Kawasaki)	0	1.7	
		30	30	93
	Tap water (Fujisawa)	0	2.5	
		30	32	98
	Rainwater (Kawasaki)	0	7.9	
		30	37	97
	River water (Okutama)	0	4.4	
		30	33	96
	Mineral drinking water (Brand A)	0	1.6	
		30	32	100
	Mineral drinking water (Brand B)	0	1.4	
		30	31	99
Bi	Tap water (Kawasaki)	0	<0.80*	
		20	20	99
	Tap water (Fujisawa)	0	<0.80*	
		20	19	97
	Rainwater (Kawasaki)	0	<0.80*	
		20	20	99
	River water (Okutama)	0	<0.80*	
		20	19	95
	Mineral drinking water (Brand A)	0	<0.80*	
		20	19	96
	Mineral drinking water (Brand B)	0	<0.80*	
		20	20	98

Sample volume: 100 mL. *Below the detection limit.

REFERENCES

1. V. Camel, "Solid Phase Extraction of Trace Elements," Spectrochimica Acta Part B: Atomic Spectroscopy, Vol. 58B, No. 7, 2003, pp. 1177-1233.

2. T. Miura, T. Morimoto, K. Hayano and T. Kishimoto, "Determination of Uranium in Water Samples by ICPAES with Chelating Resin Disk Preconcentration," Bunseki Kagaku, Vol. 49, No. 4, 2000, pp. 245-249.

3. Y. Takaku, Y. Kudo, J. Kimura, T. Hayashi, I. Ota, H. Hasegawa and S. Ueda, "Determination of Rare-Earth Elements in Environmental Water Using Chelating Resin Disk Separation/ICP-MS," Bunseki Kagaku, Vol. 51, No. 7, 2002, pp. 539-544.

4. Y. Koike, J. Sato and T. Nakamura, "Gamma-Ray Spectrometry of ^{210}Pb in Rainwater Using Preconcentration with Iminodiacetate Extraction Disk," Bunseki Kagaku, Vol. 53, No. 12, 2004, pp. 1469-1473.

5. T. Inui, K. Fujita, M. Kitano and T. Nakamura, "Determination of Cr(III) and Cr(VI) at Sub-ppb Levels in Water with Solid-Phase Extraction/Metal Furnace Atomic Absorption Spectrometry," Analytical Sciences, Vol. 26, No. 10, 2010, pp. 1093-1098.

6. T. Inui, W. Abe, M. Kitano and T. Nakamura, "Determination of Cr(III) and Cr(VI) in Water by WavelengthDispersive X-Ray Fluorescence Spectrometry after Preconcentration with an Ion-Exchange Resin Disk," X-Ray Spectrometry, Vol. 40, No. 4. 2011, pp. 301-305.

7. K. Kuriyama and Y. Kashihara, "Application of DiskFormat Solid Phase Extraction to Pesticide Analysis," Kankyo Kagaku, Vol. 5, No. 4. 1995, pp. 807-819.

8. C. Fujimoto and T. Yoshizawa, "Determination of Organochlorine Pesticides in Water Using Solid-Phase Extraction," Kankyo Kagaku, Vol. 9, No. 1, 1999, pp. 11-21.

9. Tanabe, T. Ibaraki, H. Yagoh, K. Kawata and M. Sakai, "Determination of Pesticides and Transformation Products in Environmental Water by Using a Membrane Extraction Disk," Kankyo Kagaku, Vol. 9, No. 3, 1999, pp. 607-615.

10. O. Kiguchi and K. Saitoh, "Simultaneous Determination of Polar and Nonpolar Pesticides in Water Samples by Solid-Phase Multilayered Disk Extraction Using Polystylene Gel and Activated Carbon Disk along with GC/MS," Bunseki Kagaku, Vol. 49, No. 8, 2000, pp. 575-582.

11. T. Wakui, H. Imura and K. Ohashi, "Solid-Phase Extraction and Concentration of Trace Chlorophenols with Extractant-Impregnated Membrane Disks," Bunseki Kagaku, Vol. 52, No. 9, 2003, pp. 825-828.

12. K. Kuriyama, T. Ouyang, N. Wang and Y. Furusho, "Rapid Extraction of Heavy Metals from Water Using Iminodiacetate Chelating Membrane," Kogyo Yosui, Vol. 481, No. 11, 1998, pp. 29-36.

13. T. Ouyang, N. Wang, K. Iwashima, K. Kuriyama and Y. Furusho, "Simultaneous Multi-Element Preconcentration in Aqueous Environmental Samples Using Iminodiacetate Extraction Disk (IED)

Followed by ICP-AES Analysis," Kankyo Kagaku, Vol. 9, No. 2, 1999, pp. 347-357.

14. K.-H. Lee, M. Oshima and S. Motomizu, "Inductively Coupled Plasma Mass Spectrometric Determination of Heavy Metals in Sea-Water Samples after Pre-Treatment with a Chelating Resin Disk by an On-Line Flow Injection Method," Analyst, Vol. 127, No. 6, 2002, pp. 769- 774.

15. N. E. Schmidt and S. R. Goode, "Analysis of Aqueous Solutions by Laser-Induced Breakdown Spectroscopy of Ion Exchange Membranes," Applied Spectroscopy, Vol. 56, No. 3, 2002, pp. 370-374.

16. C. R. Dockery, J. E. Pender and S. R. Goode, "Speciation of Chromium via Laser-Induced Breakdown Spectroscopy of Ion Exchange Polymer Membranes," Applied Spectroscopy, Vol. 59, No. 2, 2005, pp. 252-257.

17. X. Hou, H. L. Peters, Z. Yang, K. A. Wagner, J. D. Batchelor, M. M. Daniel and B. T. Jones, "Determination of Trace Metals in Drinking Water Using Solid-Phase Extraction Disks and X-Ray Fluorescence Spectrometry," Applied Spectroscopy, Vol. 57, No. 3, 2003, pp. 338-342.

18. W. Abe, S. Isaka, Y. Koike, K. Nakano, K. Fujita and T. Nakamura, "X-Ray Fluorescence Analysis of Trace Metals in Environmental Water Using Preconcentration with an Iminodiacetate Extraction Disk," X-Ray Spectrometry, Vol. 35, No. 3, 2006, pp. 184-189.

19. M. Resano, F. Vanhaecke and M. T. C. de Loos-Vollebregt, "Electrothermal Vaporization for Sample Introduction in Atomic Absorption, Atomic Emission and Plasma Mass Spectrometry a Critical Review with Focus on Solid Sampling and Slurry Analysis," Journal of Analytical Atomic Spectrometry, Vol. 23, No. 11, 2008, pp. 1450-1475.

20. M. Yamaguchi, N. Okubo, H. Yokoyama, T. Ui and E. Asada, "Determination of Lead in Airborne Particulates Using Direct Introduction of Filter Paper by Graphite Furnace Atomic Absorption Spectrometry," Bunseki Kagaku, Vol. 31, No. 10, 1982, pp. 561-564.

21. K. Nakano and S. Ichisugi, "Atomic Absorption Spectrometry of Silver Collected on Ion-Exchange Filter Paper by Directly Heating

the Filter Paper," Bunseki Kagaku, Vol. 34, No. 9, 1985, pp. 573-575.

22. H. Kawai, Y. Katayama, Y. Ninomiya and J. Okuda, "Determination of Trace Metals in Polyethylene Terephtalate Films by Punched-Out Film Introduction/Graphite Furnace AAS," Bunseki Kagaku, Vol. 43, No. 12, 1994, pp. 1193-1196.

23. Isozaki, Y. Morita, T. Okutani and T. Matsumura, "Direct Determination of Lead in Suspended Particulate Matter in Air by Tungsten Furnace-Atomic Absorption Spectrometry with a Punching Technique," Analytical Sciences, Vol. 12, No. 10, 1996, pp. 755-759.

24. T. Nakamura, K. Okubo and J. Sato, "Atomic Absorption Spectrometric Determination of Copper in Calcium Carbonate Scale and Carbonate Rocks by Direct Atomization of Solid Samples," Analytica Chimica Acta, Vol. 209, No. 1, 1988, pp. 287-292.

25. T. Nakamura, H. Oka, H. Morikawa and J. Sato, "Determination of Lithium, Beryllium, Cobalt, Nickel, Copper, Rubidium, Caesium, Lead and Bismuth in Silicate Rocks by Direct Atomization Atomic Absorption Spectrometry," Analyst, Vol. 117, No. 2, 1992, pp. 131-135.

26. T. Nakamura, R. Sasagawa and J. Sato, "Determination of Manganese and Copper in Synthetic and Natural Quartz Samples by Direct Atomization-AAS," Bunseki Kagaku, Vol. 41, No. 2, 1992, pp. 89-94.

27. T. Nakamura, H. Oka, M. Ishii and J. Sato, "Direct Atomization Atomic Absorption Spectrometric Determination of Be, Cr, Fe, Co, Ni, Cu, Cd, and Pb in Water with Zirconium Hydroxide Coprecipitation," Analyst, Vol. 119, No. 6, 1994, pp. 1397-1401.

28. T. Nakamura, Y. Noike, Y. Koizumi and J. Sato, "Atomic Absorption Spectrometric Determination of Copper and Lead in Silicon Nitride and Silicon Carbide by Direct Atomization," Analyst, Vol. 120, No. 1, 1995, pp. 89-94.

29. T. Nakamura, T. Kusata, H. Matsumoto and J. Sato, "Atomic Absorption Spectrometric Determination of Cadmium and Lead in Human and Artificial Teeth by Direct Atomization Technique," Analytical Biochemistry, Vol. 226, No. 2, 1995, pp. 256-262.

30. K. Fujita and T. Nakamura, "Screening of Heavy Metals in Soil by Direct Sampling/Metal Furnace Atomic Absorption Spectrometry," Bunseki Kagaku, Vol. 58, No. 7, 2009, pp. 609-615.

31. K. Matsumoto, "Palladium as a Matrix Modifier in Graphite-Furnace Atomic Absorption Spectrometry of Group IIIB-VIB Elements," Analytical Sciences, Vol. 9, No. 4, 1993, pp. 447-453.

32. F. J. Fernandez, S. A. Myers and W. Slavln, "Background Correction in Atomic Absorption Utilizing the Zeeman Effect," Analytical Chemistry, Vol. 52, No. 4, 1980, pp. 741-746.

33. Komatsu, M. Kitano and T. Nakamura, "Determination of Hazardous Metals in Plastics by Metal Furnace Atomic Absorption Spectrometry with Direct Solid-Sampling Technic," Bunseki Kagaku, Vol. 58, No. 9, 2009, pp. 785-791.

34. S. Morimoto, T. Ashino and K. Wagatsuma, "Role of a Binary Metallic Modifier in the Determination of Cadmium in Graphite Furnace Atomic Absorption Spectrometry," Analytical Sciences, Vol. 26, No. 7, 2010, pp. 809- 813.

CHAPTER 9

Photothermal cantilever deflection spectroscopy

Seonghwan Kim1, Dongkyu Lee2 and Thomas Thundat2*

1Department of Mechanical and Manufacturing Engineering, University of Calgary, T2N 1N4 Calgary, AB, Canada.

2Department of Chemical and Materials Engineering, University of Alberta, T6G 2V4 Edmonton, AB, Canada

ABSTRACT

Microcantilever sensors offer high sensitivity in the detection of adsorbed molecules based either on resonance frequency shift or changes in cantilever deflection, as both of these signals can be detected with very high resolution. Despite the high sensitivity offered by this platform, cantilevers suffer from poor selectivity due to the lack of sufficiently selective interfacial layers which can be immobilized on cantilever surfaces. This problem can be overcome by using photothermal cantilever deflection spectroscopy (PCDS), which exploits the high thermomechanical sensitivity of bi-material microcantilevers. A bi-material cantilever responds to heat generated by the nonradiative decay process when the adsorbed molecules are resonantly excited with infrared (IR) light. The variation in the cantilever deflection as a function of illuminating IR wavelength corresponds to the conventional IR absorption spectrum of the adsorbed molecules. In addition, the mass of the adsorbed molecules can be determined by measuring the resonance frequency shift of the cantilever as an orthogonal signal for the quantitative analysis. This multi-modal PCDS offers unprecedented opportunities for obtaining very high selectivity in chemical and biological sensing without using selective interfacial layers or extrinsic labels.

KEYWORDS

Bi-material microcantilever, Photothermal cantilever, deflection spectroscopy, Nanomechanical, IR spectrum Explosives, Bitumen Naphtha, DNA

1. INTRODUCTION

Recently, microcantilever sensors have attracted much attention due to their extremely high sensitivity [1],[2]. These cantilevers, which can be microfabricated into arrays using conventional micromachining techniques, offer a miniature sensing platform for the real-time, simultaneous detection of multiple target analytes using a single device [3],[4]. Another attractive feature of cantilever sensors is they can be operated in multiple modes. For example in the static mode, when confined to a single side of the cantilever, molecular adsorption results in cantilever deflection due to adsorption-induced forces [5]. In the dynamic mode, the resonance frequency of the cantilever varies sensitively as a function of adsorbed mass [5],[6].

1. 1 Molecular adsorption-induced resonance frequency variation approach

The resonance frequency, f, of a vibrating cantilever can be expressed as

$$f=\frac{1}{2\pi}\sqrt{\frac{k}{m^*}} \tag{1}$$

where k is the spring constant and m^* is the effective mass of the microcantilever. The effective mass can be related to the mass of the beam, m_b, through the relation, $m^* = n\, m_b$ where n is a geometric parameter. For a rectangular cantilever, n is 0.24. It is also possible that chemisorption or chemical reaction of the adsorbed ions/molecules may alter the spring constant of the cantilever [7]. Therefore, from equation 1 it is clear the resonance frequency can vary due to changes in mass as well as changes in spring constant. The variation in the resonance frequency, therefore, can be generalized as:

$$df=\frac{f}{2}\left[\frac{dk}{k}-\frac{dm^*}{m^*}\right] \tag{2}$$

The change in spring constant can be a result of changes in Young's modulus, E, or to changes in dimensions caused by the molecular absorption-induced swelling of the chemoselective interfacial layer such as a polymer film on the cantilever [8]. If the molecular adsorption can be confined to the terminal end of the cantilever (end loading), the contribution from the spring constant variation can be neglected. In that case, equation 1 is valid. Assuming the target molecules are evenly adsorbed on the cantilever surface and do not affect the stiffness of the cantilever, the adsorbed mass of the target molecules in air, δm, can be calculated from the simple equation:

$$\delta m=\frac{k}{(2\pi)^2 n}\left(\frac{1}{f_1^2}-\frac{1}{f_0^2}\right) \tag{3}$$

where, f_1 is the measured resonance frequency with adsorbed molecules, and f_0 is the initial resonance frequency of the cantilever [5].

1.2 Molecular adsorption-induced cantilever deflection approach

Thin microcantilevers undergo bending as a result of the mechanical forces involved in molecular adsorption. Adsorption-induced surface stress and the radius of the curvature of a cantilever can be related through Stoney's formula and the differential surface stress created by molecular adsorption can be related with the cantilever deflection as:

$$\Delta z = \frac{3L^2(1-v)}{Et^2}\delta\sigma \tag{4}$$

where L and t are the length and the thickness of the cantilever, respectively, v is the Poisson's ratio of the cantilever, and $\delta\sigma$ is the differential surface stress between the functionalized and the passivated surfaces. Therefore, the deflection of the cantilever is directly proportional to the adsorption-induced surface stress [5]. This surface stress is expressed in units of N/m or J/m^2.

For either mass or surface stress based detection, the chemical selectivity has been obtained by immobilizing chemically selective interfaces on the cantilever surface. However, this method offers only limited selectivity since most chemical interfaces based on reversible chemical interaction are not very selective [9]–[13]. In spite of recent advances in microfabricated cantilever sensors with extremely high sensitivity, most sensing applications are hampered by poor selectivity. This challenge can be traced back to fundamental limitations imposed by the chemistry of the molecular interactions which forms the basis for signal generation in currently used chemical sensors. Simple chemical interactions such as hydrogen bonding are far too general for providing selectivity. In addition, immobilized chemoselective interfaces have a limited shelf-life as they degrade over time resulting even poorer selectivity [9].

We identify the lack of chemical selectivity in small molecule detection as the main challenge in accepting microcantilever sensors in practical applications. The limited selectivity and sensitivity of chemical sensors stems from the fundamental concept involved in the current sensing process which operates by the "molecule detects molecule" principle of selectivity. Adsorption of analyte molecules on the chemoselective interface immobilized on the cantilever surface results in variations in its mass and surface stress. Many current nano and micro sensors suffer from similar selectivity challenge as they are based on monitoring the changes in one of its physical properties, such as adsorbed mass, temperature, surface stress, resistance, capacitance, or refractive index, due to molecular adsorption. The transducer response is then amplified for readout and display. Despite all the advances in micro/nanotechnology, the fundamental mechanism of selectivity in sensing has not yet changed. In fact, current advances in chemical sensors are focused on developing transducers, which can detect extremely small changes in physical properties as a result of molecular adsorption. As long as the fundamental working principle for signal generation remains the same, the advantages of micro/nanotechnology will remain untouched in overcoming the challenges of sensor selectivity. Without selectivity, all other advantages (high sensitivity, fast response time, decrease in size and power consumption, and the potential for simultaneous detection of multiple analytes) are of no practical use.

Designing highly selective molecular recognition layers that can be immobilized on a sensor surface for small molecule detection is a challenging task due to the limited number of chemical interactions that can serve as the basis for synthesizing chemoselective layers while satisfying the highly desirable sensor attribute of room temperature reversibility. One way of achieving selectivity in detection is by using arrays of cantilevers where each element is modified with partially selective interfaces [10]–[12]. The requirement of room temperature reversibility requires the use of weak chemical interactions, for example hydrogen bonding, between the target molecules and the chemoselective interfaces. If the response from each cantilever is unique for a given chemical, it is possible to analyze the response using pattern recognition algorithms. The responses can include rise and decay times, as well as amplitudes. However, molecular recognition based on chemical interfaces relying on weak interactions is not specific enough to produce unique response pattern even in an array format. Increasing the number of sensor elements in the array for pattern recognition analysis does not improve the selectivity [13].

Imparting chemical selectivity to cantilever sensors is a challenge and will require utilizing unique properties of cantilever sensors in order to develop an entirely novel approach. One of the most under exploited properties of a cantilever sensor is its extremely high sensitivity to temperature variation when it is fabricated as a bi-material beam. This extreme high thermomechanical sensitivity of a bi-material cantilever can be exploited for measuring the extremely small thermal changes due to nonradiative decay of molecular vibrations of the adsorbed molecules [14]–[25]. The molecular vibrations of the adsorbed molecules can be resonantly excited by illuminating using infrared (IR) light. This change in heat energy due to molecular excitations on the cantilever sensor results in cantilever deflection. The amplitude of the bi-material cantilever deflection as a function of an illuminating wavelength shows the amount of heat energy generated by the adsorbed molecules at that wavelength and matches very well with the IR absorption peaks of the adsorbed molecules. This photothermal cantilever deflection spectroscopy (PCDS) is a unique technique that combines the extremely high thermomechanical sensitivity of a bi-material cantilever with the selectivity of mid-IR spectroscopy to achieve selectivity and sensitivity in molecular recognition of adsorbed molecules. By observing the energy location of multiple peaks in the cantilever deflection, it is possible to identify different molecules.

In this review article, we describe the working principle of PCDS, its instrumentation and methods. Quantitative results from the label-free, receptor-free detection of many analytes are presented. We also discuss potential strategies to enhance the thermomechanical sensitivity of bi-material cantilevers.

2. REVIEW

The PCDS setup and its working principle are shown in Figure 1. Vapor phase analyte molecules are first allowed to adsorb on the bi-material cantilever surface. The mass of the adsorbed molecules is then calculated from the initial and final resonance frequencies of the cantilever. The cantilever is then illuminated with IR light from either a monochromator with a glow bar and a filter wheel or a broadly tunable quantum cascade laser (QCL). When IR photons of specific wavelength are resonantly absorbed by the adsorbates on the bi-material cantilever, it results in the generation of heat causing instantaneous cantilever deflection. The cantilever bends in response to this variation in temperature. By using a differential technique (cantilever with and without adsorbed molecules), it is possible to obtain a nanomechanical IR absorption spectrum of the adsorbates by plotting the differential cantilever deflection as a function of illumination wavelength. This signal depends on the thermodynamic and energy transfer properties of the adsorbates and the cantilever beam. Temperature changes resulting from the absorption of IR energy are directly related to the vibrational modes of the adsorbed molecules as well as to the heat capacity and thermal conductivity of the cantilever beam. The observed peak amplitudes of nanomechanical IR spectra are proportional to the amount of the adsorbed molecules, the impinging power of IR radiation, the absorption mode, and the thermomechanical sensitivity of the bi-material cantilever. Since the PCDS is based on nonradiative decay, it does not rely on Beer-Lambert's principle. Unlike, techniques based on Beer-Lambert's principle, such as Fourier transform infrared (FTIR) and conventional IR spectroscopy, the signal-to-noise ratio of PCDS increases with the power of the illumination source. Therefore, by increasing the power of the IR light source, it is possible to increase the sensitivity of the technique [21].

Since the resonance frequency of the cantilever can be monitored before and after molecular adsorption, the adsorbed mass can be simultaneously determined using equation 3. Therefore, within a certain dynamic range, the peak amplitudes of the nanomechanical IR spectra can be normalized by adsorbed mass and these can serve as basis for quantitative nanomechanical IR spectral analysis to determine the adsorbed mass of each target molecule in a mixture since the IR spectrum of a mixture is a linear superposition of individual spectra.

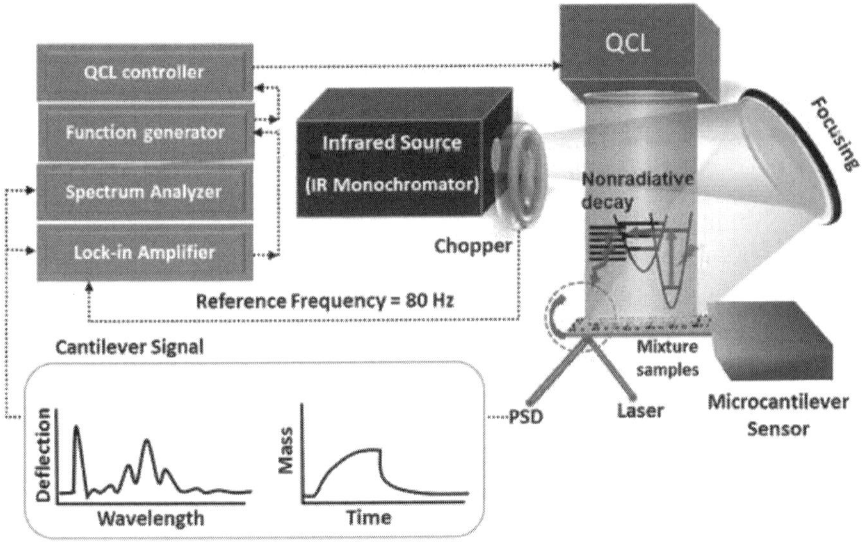

Figure 1. Schematic illustration of the experimental setup. Orthogonal signals (nanomechanical IR spectra and mass variations) are measured by optical beam deflection method using a red diode laser and a position sensitive detector (PSD). These nonradiative decay processes result in heating up the bi-material cantilever, generating the deflection of the cantilever.

Since IR spectra are different for different molecules, high chemical selectivity can be achieved in detection. This technique can be used for selective detection and quantification of the binary and ternary mixtures of explosive molecules such as trinitrotoluene (TNT), cyclotrimethylene trinitramine (RDX), and pentaerythritol tetranitrate (PETN)) [21]. Figure 2 shows the normalized nanomechanical IR absorption spectra of explosive molecules. Figure 2a presents the normalized IR spectra of 1:1 binary mixtures such as TNT&RDX (blue), PETN&TNT (cyan), and RDX&PETN (magenta). The nanomechanical IR spectra of the individual explosives were taken separately as references and agreed quite well with our previous report [20]. Figure 2b shows the normalized nanomechanical IR absorption spectra of individual explosive molecules, TNT (black), PETN (red), and RDX (green), using a conventional monochromator. The signal amplitude is in the range of mV/ng. However, the amplitude of normalized nanomechanical IR absorption spectra for the same samples using the QCL in Figure 2c were three orders of magnitude higher than that of spectra using a monochromator in Figure 2b. The sensitivity of PCDS was improved by employing a broadly tunable QCL as a powerful IR source [21]. Several distinct peaks and shoulders appeared in the binary mixture spectrum since the mixture spectrum is a linear superposition of individual spectra. Comparing these peaks with those of individual TNT, RDX, and PETN spectra, it is apparent that the peaks at 6.49 and 7.46 μm are from TNT, the peaks at 6.38 and 7.27 μm are from RDX, and the peaks at 6.06 and 7.82 μm are from PETN molecules [26]–[30]. PCDS can detect and distinguish differences between such closely related molecular species and between other interfering compounds and target molecules while sensing in mixture samples.

Selective and sensitive detection has direct relevance in many industrial applications such as chemical process and environmental monitoring [22],[31]. For example, naphtha - a complex hydrocarbon – is routinely used for extraction of bitumen from oil sands, needs to be monitored on a routine basis in oil sands processing. Since the process of stripping naphtha from bitumen is not perfect, they easily lost into tailing streams during bitumen extraction process. However, the detection of industrial solvent and chemicals such as high volatile organic compounds poses many challenges as a result of non-specific interactions with surfaces and interferences from identical groups in other compounds. By monitoring multiple peaks of adsorbed molecules in PCDS, it is possible to overcome the selectivity and sensitivity challenges of currently used techniques. Figure 3 demonstrates selective detection of naphtha and bitumen using PCDS. The IR spectra of bitumen and naphtha reveal that they are largely composed of the aliphatic compounds. The common bands correspond to C–H asymmetrical stretching of $-CH_2$ and $-CH_3$ in range of 3.35 to 3.5 μm. The two bands at 6.8 and 7.2 μm are associated with C–H deformation in $-CH_2$ and $-CH_3$, respectively. Carbon rings in a cyclic compound and CH out of plane deformations in vinyl compounds show peaks in the range of 9.7–10.3 μm [22],[31]. However, we can clearly distinguish them with some characteristic peaks, which can be attributed to different ratio of each component and some minor mixture components having various functional groups such as carboxylic acids (C–O, C–O–H) and ring vibrations (C = C, C–C).

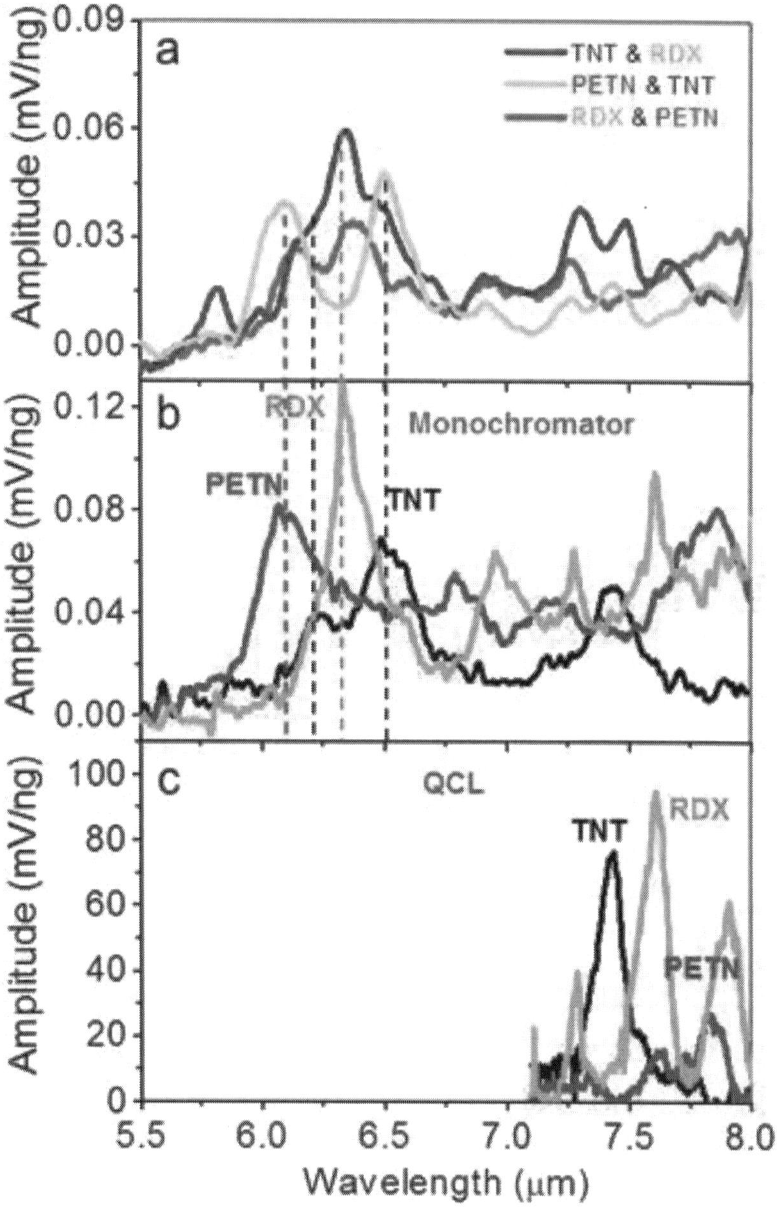

Figure 2. Normalized nanomechanical IR absorption spectra of explosive molecules. **(a)** Normalized IR spectra of 1:1 binary mixtures (TNT&RDX (blue), PETN&TNT (cyan), RDX&PETN (magenta)), and three individual explosive component (TNT (black), PETN (red), RDX (green)) using **(b)** monochromator-based and **(c)** QCL-based PCDS.

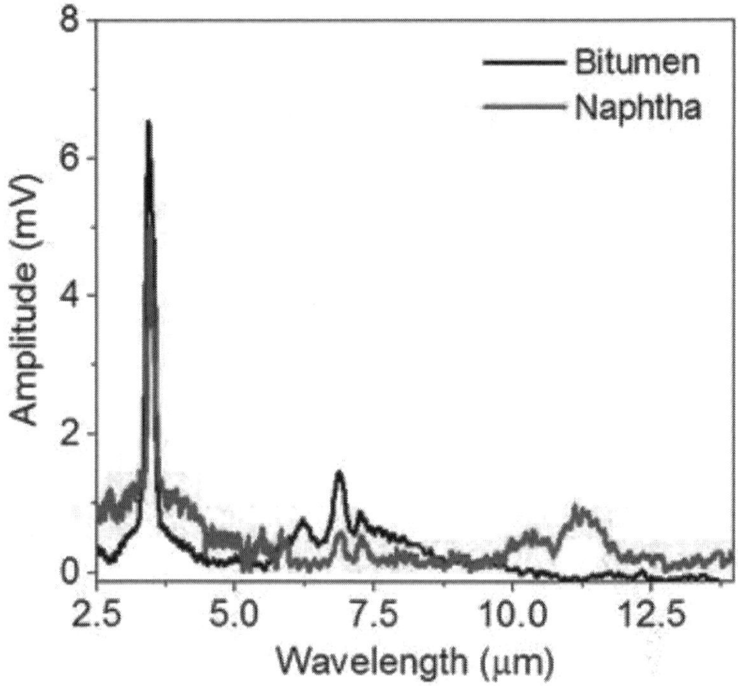

Figure 3. Nanomechanical IR absorption spectra of bitumen (black) and naphtha (red).

The PCDS technique also offers high selectivity in detection of biological samples. Despite intrinsic selectivity in the binding of antigen-antibody or DNA hybridization, the microfabricated biosensors incur several problems such as non-uniform immobilization, reduced activity of receptors, and non-specific binding of unwanted molecules. Additionally, it is not easy to rapidly detect biological target molecules due to the intensive sample delivery process and reaction time involved in the receptor-target interactions. To overcome these issues, the PCDS technique can be used as an effective label-free and receptor-free biosensing concept. Figure 4 shows the normalized nanomechanical IR absorption spectra of the ssDNA of 20 bases of guanine using a conventional monochromator (Figure 4a). Figure 4b shows the same ssDNA spectrum obtained using a QCL which can scan from 5.68 μm to 6.39 μm. In Figure 4, the peaks between 5.68 and 6.39 μm result from the stretching vibration of the carbonyl group and ring bonds in guanine.

The peaks of guanine are from carbonyl C = O displacement (5.91 μm) and ring bond C = N stretching (6.1, 6.25 μm) [32],[33]. The sensitivity of QCL-based PCDS was found to be superior to that of monochromator-based PCDS as a result of using a high power IR source. These experiments show that QCL-based PCDS has the great potential for rapid identification and quantification of biological analytes. Although this technique is not able to discern different DNA sequences, it is highly useful as a rapid screening method for potential variation (mutation) in a given DNA strand with a known-sequence as well as detection of conjugation and byproduct of biomolecular interactions [32],[34],[35]. This technique may also find forensic applications where the need is to discriminate DNA rather than sequencing.

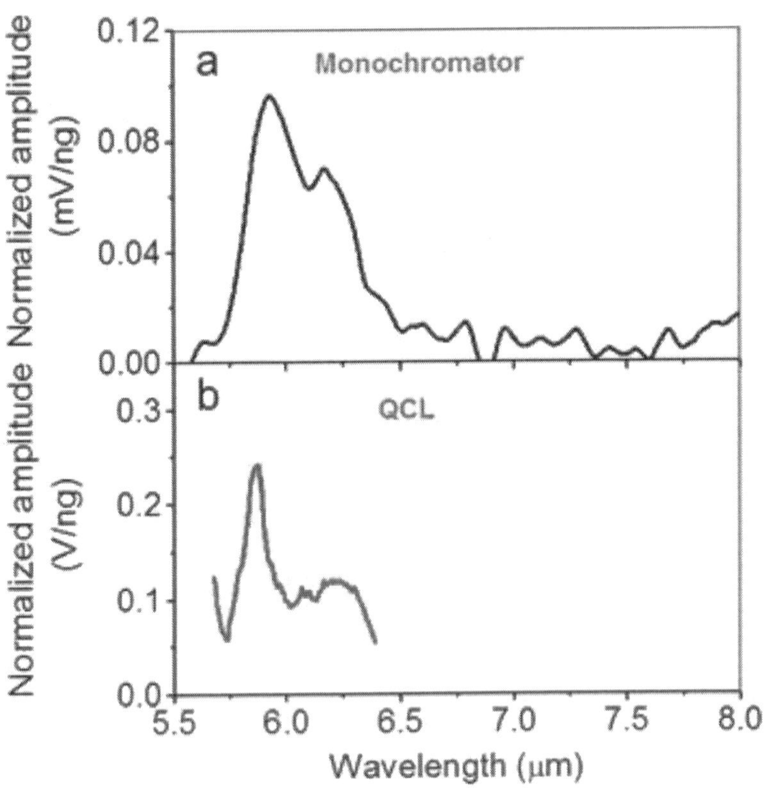

Figure 4. Normalized nanomechanical IR absorption spectra of ssDNA of 20-mer guanine using monochromator-based (a) and (b) QCL-based PCDS.

The relative intensities of the PCDS peaks were different than those observed with conventional IR spectrum or FTIR method. Some peaks that are not very prominent in conventional IR spectra appear to have higher intensity in the PCDS spectra. This may be directly related to the efficiency of nonradiative decay of these excited states. The position of absorption peaks were in general agreement with those observed with conventional IR absorption spectroscopies. We did not observe any appreciable shift in the position of the PCDS peaks. Since the energy resolution of the PCDS depends on the energy resolution of the IR sources, any small shift induced by the presence of the substrate is below the resolution limit of the current device.

3. ADVANTAGES OF THE PCDS TECHNIQUE

The PCDS does not use any chemical receptors or interfaces for molecular recognition. The cantilever used in this technique has no surface functional groups. The molecules can be deposited on the surface with no chemical interaction between the analyte and the substrate. Even placing a drop of solution and drying it under ambient conditions is sufficient to obtain PCDS signals.

- High Selectivity: The signal generation is based on resonance excitation of various molecular bonds in the analyte. Molecular vibrations are extremely selective. In addition, in a complex mixture, the resultant signal is a linear combination of molecular vibrations from various constituents. Therefore, using a simple superposition principle it is possible to analyze the signal for the target analyte.

- High Sensitivity: The PCDS combines the extreme high thermomechanical sensitivity of a bi-material cantilever beam with the selectivity of mid-IR molecular spectroscopy. Our earlier results have shown mass sensitivity of 10 pg on the cantilever (corresponding to a few monolayers of analyte) under ambient conditions.

- Quantitative: The magnitude of the IR absorption-induced cantilever bending is proportional to the amount of analyte on the cantilever. This can be further quantified by measuring the resonance frequency of the cantilever (mass loading).

- Real-time Monitoring: Since the heat generation due to non-radiative decay is instantaneous, the PCDS technique is capable of real-time monitoring of adsorbed analytes.

- Small Size: The first generation device can be packaged to be smaller than a shoe box (including IR source, power supplies, display unit, etc.) making it a portable device. A second generation device can be made even smaller (the size of a cell phone) by optimizing packaging.

- Low Cost: Once produced in bulk, the cost of the device will be a few thousand dollars.

- Versatility: The same device can be used for different analytes.

4. CONCLUSIONS

PCDS provides simple technique to impart chemical selectivity onto microcantilever sensors. The demonstrated sensitivity and selectivity of this approach offers new possibilities for receptor-free sensing of a wide range of materials beyond what is currently possible using conventional techniques. These devices have the obvious advantages of requiring sub-nanogram samples, quick detection time and the potential to be inexpensive. It is possible to increase the sensitivity of PCDS by optimizing the bi-material cantilever parameters as well as increasing the power of the illuminating IR source. For example, by selecting different metals and optimizing the thickness of the coating it is possible to make a bi-material cantilever very sensitive to thermal changes. It is also possible to fabricate cantilevers with an optimized spring constant for increased thermomechanical bending or to pattern the cantilever surface for increased adsorption. Additionally, sensitivity can be increased by restricting the heat flow from the cantilever onto the base of the cantilever by making the contact area smaller.

5. METHOD

5.1 Maximizing the thermomechanical sensitivity of a bi-material cantilever

A microcantilever which is made bi-material by depositing a thin layer of metal on one of its side becomes an excellent, miniature thermal sensor, capable of detecting extremely small temperature changes at room temperature (there is no need to cool the device). The deflection of the cantilever tip due to the bi-material effect can be found from the following equation [36]:

$$z = -\frac{3}{4}(\alpha_1 - \alpha_2)\frac{t_1 + t_2}{t_2^2 K}\frac{l^3}{(\lambda_1 t_1 + \lambda_2 t_2)w}P. \tag{5}$$

where z is the cantilever deflection at the tip, α_1 and α_2 are the coefficients of thermal expansion for the two layers, l is the length of the cantilever, t_1 and t_2 are the layer thicknesses, λ_1 and λ_2 are the thermal conductivities, w is the width of the cantilever, and P is the total power absorbed by the cantilever. The subscripts 1 and 2 represent metal and cantilever substrate, respectively. The parameter K stands for the expression:

$$K = 4 + 6\left(\frac{t_1}{t_2}\right) + 4\left(\frac{t_1}{t_2}\right)^2 + \frac{E_1}{E_2}\left(\frac{t_1}{t_2}\right)^3 + \frac{E_2}{E_1}\left(\frac{t_2}{t_1}\right). \tag{6}$$

In the PCDS technique, the cantilever with adsorbed chemical species is scanned using successive pulses of monochromatic IR radiation. The amplitude of cantilever deflection as a function of IR wavelength resembles the IR absorption spectrum of the adsorbed molecules. The limit of detection of this technique is determined by the power of IR source and the thermomechanical sensitivity of the cantilever, which can be optimized by properly designing the bi-material aspect of the cantilever.

5.2 Instrumentation

5.2.1 Preparation of bi-material cantilevers

The experiments were carried out using commercially available silicon microcantilevers. Rectangular silicon cantilevers (CSC12-E) were obtained from MikroMasch USA (San Jose, CA). The dimension of each cantilever was 350 μm in length, 35 μm in width, and 1 μm in thickness. Another rectangular silicon cantilevers (Octosensis) were obtained from Micromotive GmbH Germany. A silicon chip contained an array of eight cantilevers having typical dimensions of 500 μm in length, 90 μm in width, and 1.0 ± 0.3 μm in thickness. The microcantilevers were cleaned by rinsing with acetone, ethanol, and treating them with UV ozone and then coated with 10 nm of chromium (adhesion layer) followed by 200 nm of gold using an e-beam evaporator at the deposition rate of 0.2 Å/s. A bare Octosensis silicon cantilever was used for bitumen experiments. Bitumen was diluted with toluene and coated on top of the silicon cantilever using a micro glass capillary. The bitumen-coated cantilever was dried in a vacuum oven at 150°C overnight to remove any low molecular weight solvent mixed in with the bitumen. Three standard explosive samples (TNT, RDX, and PETN) were purchased from AccuStandard, Inc. (New Haven, CT) and used without further purification. As indicated by the manufacturer, the standard concentration of each explosive is 1 mg/mL. The binary mixtures (by volume of standard sample solution) of explosive molecules were deposited on CSC12-E microcantilevers using micro glass capillaries. Naphtha was deposited on the gold side of the Octosensis cantilever by drop casting. The used chemicals (acetone, ethanol, toluene, and naphtha) were purchased from Fisher Scientific and used without further purification. We also procured synthetic single-stranded DNA (ssDNA) of 20 bases of guanine (G20) from Oligos Etc. The initial concentration of the ssDNA solution was 10 μM in deionized water. DNAs were deposited on the cantilever by inserting the cantilever into a micro glass capillary containing the DNA solution and slowly taking the cantilever out of the capillary.

5.2.2 PCDS setup

The sample-coated cantilever was mounted on a stainless steel cantilever holder and then positioned in the head unit of a Multi-Mode atomic force microscope (AFM) (Bruker, Santa Barbara, CA). The nanomechanical deflection and resonance frequency of the cantilever were measured using the optical beam deflection method with a red laser diode and a position sensitive detector (PSD). The IR radiation from the monochromator (Foxboro Miran 1A-CVF) was mechanically chopped at a frequency of 80 Hz and focused onto the cantilever by a spherical mirror. The filter wheel of the monochromator scans from 2.5 μm to 14.5 μm (i.e., 4000 cm^{-1} to 690 cm^{-1} in wavenumber) and has a spectral resolution of approximately 0.05 μm at 3 μm, 0.12 μm at 6 μm, and 0.25 μm at 11 μm according to the manufacturer. The 200 kHz pulsed IR radiation with a 10% duty cycle from a ÜT-8 QCL (Daylight Solutions, San Diego, CA) was electrically modulated at 80 Hz, using a DS345 function generator (Stanford Research Systems, Sunnyvale, CA) and directed to the cantilever for explosive sensing experiments. The IR wavelength was scanned from 7.1 μm to 8.73 μm (1408 cm^{-1} to 1145 cm^{-1}) with a step size of 5 nm. For DNA sensing experiments, 100 kHz pulsed IR radiation with a 5% duty cycle from a QCL (Daylight Solutions 6 μm laser) was electrically modulated at 80 Hz and directed to the DNA-coated cantilever. The IR wavelength was scanned from 5.68 μm to 6.39 μm (1760 cm^{-1} to 1565 cm^{-1}) with a step size of 5 nm. The resonance frequencies of the cantilever were measured before and after sample deposition using an SR760 spectrum analyzer (Stanford Research Systems, Sunnyvale, CA). The nanomechanical IR deflection signals before and after sample deposition were taken using an SR850 lock-in amplifier (Stanford Research Systems, Sunnyvale, CA) and the differential signals were presented as the nanomechanical IR spectra.

AUTHORS' INFORMATION

Seonghwan Kim is an Assistant Professor in the Department of Mechanical and Manufacturing Engineering at the University of Calgary. He earned his B.S. and M.S. degree in aerospace engineering at Seoul National University in 1998 and 2000, respectively, and received his Ph.D. in mechanical engineering at the University of Tennessee, Knoxville in 2008.

Dongkyu Lee is a Postdoctoral fellow in the Department of Chemical and Materials Engineering, University of Alberta, Edmonton. He received his Ph.D. in chemical engineering from Pohang University of Science and Technology, Pohang, Korea in 2011.

Thomas Thundat is a Canada Excellence Research Chair Professor at the University of Alberta, Edmonton. He received his Ph.D. in physics from the State University of New York at Albany in 1987.

ACKNOWLEDGEMENTS

This work is supported by Canada Excellence Research Chairs (CERC) Program. D.L. also would like to acknowledge the partial support from Basic Science Research Program through the National Research Foundation of Korea (NRF) funded by the Ministry of Education, Science and Technology (2012R1A6A3A03040416). SK also acknowledges the partial support from the Schulich School of Engineering in the University of Calgary.

COMPETING INTERESTS

The authors declare that they have no competing interests.

AUTHORS' CONTRIBUTIONS

SK & DL designed, carried out the experiments, collected and analyzed the data, and wrote the manuscript; TT conceived the experiments, analyzed results, and wrote the manuscript. All authors discussed the results and commented on the manuscript. All authors read and approved the final manuscript.

REFERENCES

1. Thundat T, Warmack RJ, Chen GY, Allison DP: Thermal and ambient-induced deflections of scanning force microscope cantilevers. Appl Phys Lett 1994, 64: 2894–2896.

2. Gimzewski JK, Gerber C, Meyer E, Schlittler RR: Observation of a chemical reaction using a micromechanical sensor. Chem Phys Lett 1994, 217: 589–594.

3. Baller MK, Lang HP, Fritz J, Gerber C, Gimzewski JK, Drechsler U, Rothuizen H, Despont M, Vettiger P, Battiston FM, Ramseyer JP, Fornaro P, Meyer E, Güntherodt HJ: A cantilever array-based artificial nose. Ultramicroscopy 2000, 82: 1–9.

4. Fritz J, Baller MK, Lang HP, Rothuizen H, Vettiger P, Meyer E, Güntherodt HJ, Gerber C, Gimzewski JK: Translating biomolecular recognition into nanomechanics. Science 2000, 288: 316–318.

5. Chen GY, Thundat T, Wachter EA, Warmack RJ: Adsorption-induced surface stress and its effects on resonance frequency of microcantilevers. J Appl Phys 1995, 77: 3618–3622.

6. Thundat T, Chen GY, Warmack RJ, Allison DP, Wachter EA: Vapor detection using resonating microcantilevers. Anal Chem 1995, 67: 519–521.

7. Cherian S, Thundat T: Determination of adsorption-induced variation in the spring constant of a microcantilever. Appl Phys Lett 2002, 80: 2219–2221.

8. Wachter EA, Thundat T: Micromechanical sensors for chemical and physical measurements. Rev Sci Instrum 1995, 66: 3662–3667.

9. Hsieh MD, Zellers ET: Limits of recognition for simple vapor mixtures determined with a microsensor array. Anal Chem 2004, 76: 1885–1895.

10. Then D, Vidic A, Ziegler C: A highly sensitive self-oscillating cantilever array for the quantitative and qualitative analysis of organic vapor mixtures. Sens Actuators B 2006, 117: 1–9.

11. Zhao W, Pinnaduwage LA, Leis JW, Gehl AC, Allman SL, Shepp A, Mahmud KK: Identification and quantification of components in ternary vapor mixtures using a microelectromechanical-system-based electronic nose. J Appl Phys 2008, 103: 104902.

12. Jin C, Kurzawski P, Hierlemann A, Zellers ET: Evaluation of multitransducer arrays for the determination of organic vapor mixtures. Anal Chem 2008, 80: 227–236.

13. Jin C, Zellers ET: Limits of recognition for binary and ternary vapor mixtures determined with multitransducer arrays. Anal Chem 2008, 80: 7283–7293.

14. Barnes JR, Stephenson RJ, Welland ME, Gerber C, Gimzewski JK: Photothermal spectroscopy with femtojoule sensitivity using a micromechanical device. Nature 1994, 372: 79–81.

15. Wachter EA, Thundat T, Oden PI, Warmack RJ, Datskos PG, Sharp SL: Remote optical detection using microcantilevers. Rev Sci Instrum 1996, 67: 3434–3439.

16. Li G, Burggraf LW, Baker WP: Photothermal spectroscopy using multilayer cantilever for chemical detection. Appl Phys Lett 2000, 76: 1122–1124.

17. Datskos PG, Rajic S, Sepaniak MJ, Lavrik N, Tipple CA, Senesac LR, Datskou I: Chemical detection based on adsorption-induced and photoinduced stresses in microelectromechanical systems devices. J Vac Sci Technol B 2001, 19: 1173–1179.

18. Arakawa ET, Lavrik NV, Rajic S, Datskos PG: Detection and differentiation of biological species using microcalorimetric spectroscopy. Ultramicroscopy 2003, 97: 459–465.

19. Wig A, Arakawa ET, Passian A, Ferrell TL, Thundat T: Photothermal spectroscopy of Bacillus anthracis and Bacillus cereus with microcantilevers. Sens Actuators B 2006, 114: 206–211.

20. Krause AR, Van Neste C, Senesac L, Thundat T, Finot E: Trace explosive detection using photothermal deflection spectroscopy. J Appl Phys 2008, 103: 094906.

21. Kim S, Lee D, Liu X, Van Neste C, Jeon S, Thundat T: Molecular recognition using receptor-free nanomechanical infrared spectroscopy based on a quantum cascade laser. Sci Rep 2013, 3: 1111.

22. Bagheri M, Chae I, Lee D, Kim S, Thundat T: Selective detection of physisorbed hydrocarbons using photothermal cantilever deflection spectroscopy. Sens Actuators B 2014, 191: 765–769.

23. Lee D, Kim S, Van Neste CW, Lee M, Jeon S, Thundat T: Photoacoustic spectroscopy of surface adsorbed molecules using a nanostructured coupled resonator array. Nanotechnology 2014, 25: 035501.

24. Khan MF, Kim S, Lee D, Schmid S, Boisen A, Thundat T: Nanomechanical identification of liquid reagents in a microfluidic channel. Lab Chip 2014, 14: 1302–1307.

25. Lee D, Kim S, Chae I, Jeon S, Thundat T: Nanowell-patterned TiO2 microcantilevers for calorimetric chemical sensing. Appl Phys Lett 2014, 104: 141903.

26. Pristera F, Halik M, Castelli A, Fredericks W: Analysis of Explosives Using Infrared Spectroscopy. Anal Chem 1960, 32: 495–508.

27. Lewis IR, Daniel NW Jr, Griffiths PR: Interpretation of raman spectra of nitro-containing explosive materials. part i: group frequency and structural class membership. Appl Spectrosc 1997, 51: 1854–1867.

28. Makashir PS, Kurian EM: Spectroscopic and thermal studies on 2,4,6-trinitro toluene (TNT). J Therm Anal Cal 1999, 55: 173–185.

29. Makashir PS, Kurian EM: Spectroscopic and thermal studies on Pentaerythritol Tetranitrate (PETN). Propell Explos Pyrot 1999, 24: 260–265.

30. Beal RW, Brill TB: Vibrational Behavior of the –NO2 Group in Energetic Compounds. Appl Spectrosc 2005, 59: 1194–1202.

31. Yoon S, Son J, Lee W, Lee H, Lee CW: Prediction of bitumen content in oil sand based on FT-IR measurement. J Ind Eng Chem 2009, 15: 370–374.

32. Dovbeshko GI, Gridina NY, Kruglova EB, Pashchuk OP: FTIR spectroscopy studies of nucleic acid damage. Talanta 2000, 53: 233–246.

33. Kim S, Lee D, Thundat R, Bagheri M, Jeon S, Thundat T: Photothermal cantilever deflection spectroscopy. ECS Trans 2013, 50: 459–464.

34. Mateo-Marti E, Briones C, Roman E, Briand E, Pradier CM, Martin-Gago JA: Self-assembled monolayers of peptide nucleic acids on gold surfaces: a spectroscopic study. Langmuir 2005, 21: 9510–9517.

35. Liao W, Wei F, Liu D, Qian MX, Yuana G, Zhaoa XS: FTIR-ATR detection of proteins and small molecules through DNA conjugation. Sens Actuators B 2006, 114: 445–450.

36. Lai J, Perazzo T, Shi Z, Majumdar A: Optimization and performance of high-resolution micro-optomechanical thermal sensors. Sens Actuators A 1997, 58: 113–119.

Index